"Let There Be Light!"

"Let There Be Light!"

Nuclear Energy: A Christian Case

ROBERT S. DUTCH

Foreword by Kenneth Stewart

WIPF & STOCK · Eugene, Oregon

"LET THERE BE LIGHT!"
Nuclear Energy: A Christian Case

Copyright © 2017 Robert S. Dutch. All rights reserved. Except for brief quotations in critical publications or reviews, no part of this book may be reproduced in any manner without prior written permission from the publisher. Write: Permissions, Wipf and Stock Publishers, 199 W. 8th Ave., Suite 3, Eugene, OR 97401.

Wipf & Stock
An Imprint of Wipf and Stock Publishers
199 W. 8th Ave., Suite 3
Eugene, OR 97401

www.wipfandstock.com

PAPERBACK ISBN: 978-1-4982-9149-1
HARDCOVER ISBN: 978-1-4982-9151-4
EBOOK ISBN: 978-1-4982-9150-7

Manufactured in the U.S.A.   JUNE 14, 2017

Scripture taken from the Holy Bible, New International Version®, NIV® Copyright ©1973, 1978, 1984, 2011 by Biblica, Inc.™ Used by permission. All rights reserved worldwide.

(Checked using: https://www.biblegateway.com/)

To my parents, Stan and Lil, for their love and constant support from my childhood days—growing up during the austerity years in post-war Liverpool, England.

And to Susan, my wife, our children, Laura, Jess, and Tim, and our grandchildren for their invaluable contributions, directly and indirectly, in supporting me and seeing this book published. Thinking about their futures in our world influenced my writing.

Contents

List of Tables | x
Foreword by Kenneth Stewart | xi
Preface | xiii
Acknowledgments | xv
Abbreviations | xviii
Introduction | xxv
    Electricity, Energy, and Environment
    Conclusion
    Further Reading

1   Our Nuclear and Radioactive Universe | 1
    Introduction
    From the Big Bang to Our Solar System: Origin and Structure of the Universe
    The Sun
    Stars and Supernovae: Origin of the Elements
    The Earth
    Nuclear Fission and Natural Nuclear Reactors
    Natural Radioactivity
    Conclusion
    Further Reading
    Questions

2   Radioactivity and Radiation Exposure on Earth | 16
    Introduction
    Exposure of the UK Population from Radiation in the Environment
    Exposure of the UK Population from Using Radiation
    Exposure of Ireland's Population
    Exposure of the US Population
    Further Information on Artificial (Anthropogenic) Sources
    Radioactivity in the UK's Food and Environment
    Conclusion
    Further Reading
    Questions

CONTENTS

3 Recognizing Risks and Benefits | 35

Introduction
Risks in Everyday Living
Alcohol and Smoking: Benefits and Risks
Discovery and Health
Electricity Generation
Civil Nuclear Energy and Risks
UK New Nuclear Plants: Justification for Building
Radiation Risks and Radon
Medical Radiation: Benefits and Risks
Medical CT Scans
Electricity and Health: Benefits and Risks
Nuclear Power: Professional and Public Radiation Risk Perceptions
Radioactive Waste and Low Doses
Morally Responsible Risk Communication
Conclusion
Further Reading
Questions

4 Energy Policies and Positions | 58

Introduction
Electricity Generation: The Role of Nuclear Energy
Environmentalists
UK Nuclear Policy
Political Parties
The Regulators
Universities/Colleges and Employers
Trade Unions
Churches and Christians
Environment, Energy, and Ethics
Conclusion
Further Reading
Questions

5 Nuclear Energy: Concerns, Questions, Objections (Part 1) | 102

Introduction
Public Attitudes and Understanding
Objection 1: There Is a Link between Nuclear Energy and Nuclear Weapons
Objection 2: Uranium Supplies Are Limited
Objection 3: Building Nuclear Power Plants Uses Vast Quantities of Concrete and Steel
Objection 4: Nuclear Power Plants Cannot Be Built Fast Enough to Meet Our Needs
Objection 5: Decommissioning Costs Are Too High
Conclusion
Further Reading
Questions

CONTENTS

6   Nuclear Energy: Concerns, Questions, Objections (Part 2) | 124

Introduction
Objection 6: Safety
Objection 7: Security
Objection 8: Safeguards
Objection 9: Accidents
Objection 10: Waste
Conclusion
Further Reading
Questions

7   Grasping Our Future! | 149

Introduction
Christians, Climate Change, and Paris COP21
Skills Shortages
Radioactive Waste and Research into the Fuel Cycle
Fast Reactors
"Too Cheap to Meter"
Nuclear Energy Post-Fukushima
A Future with Renewables and Reactors
Thorium Reactors
Small Modular Reactors
Fusion Reactors
Nuclear Energy: The Benefits
Christians, Churches, and Christian Leaders
Local Support for Nuclear Workers
Hope
Conclusion
Further Reading
Questions

Glossary | 181

Bibliography | 183

List of Tables

Table 1   Uranium Isotopes: Half-Lives and Percentage Levels in Uranium Ore | 13
Table 2   Average Annual UK Population Exposures in 2010 | 26
Table 3   Average Annual US Population Exposures in 2006 | 28

Foreword

THE BASIC PREMISE OF this book is that for too long nuclear power has had a bad press, an arguably undeserved bad press. As a result, too many of us are guilty of uninformed prejudice against nuclear generated power, leading to the lazy assumption that it is just too expensive, too dangerous, too polluting ever to consider using. My former colleague, Robert Dutch, wants to address and answer those important concerns, and having spent most of his adult life working in the nuclear power industry within the UK, predominantly in the crucial area of nuclear safety, he writes out of close personal engagement with his subject. Yes, he is unashamedly an advocate of the use of nuclear power and argues that a proportion of our electricity must be generated from nuclear energy if we are ever to supply our projected future needs while at the same time reducing our dependence on fossil fuels with the related consequences of their usage for global warming and climate change.

Significantly, Robert doesn't only present his case on scientific and technological grounds, but also out of personal Christian conviction. Having gained theological qualifications in later life, he wishes to argue for the use of nuclear power on the basis of his faith in God, as the Creator of a universe in which nuclear fusion and fission are fundamental processes, where radioactive decay is as much a part of God's natural order as the carbon cycle or photosynthesis we learnt about in school. Nuclear power and radioactivity are not to be regarded as "works of the devil," but God-ordained processes which have gifted to us the universe we know today.

That said, in seeking to use these for the benefit of humankind, we need to recognize that they do bring risks which must be addressed responsibly. Robert will show that risk is not exclusive to the nuclear industry, but actually exists in other everyday areas of life that most of us rarely consider. To make our decisions about nuclear power, we need facts, not plucked from a headline designed to sell copy, but from the best research reports available and set in the widest context. This book sets out to provide that very resource.

It is aimed primarily not at those with a background in physics or science in general, but at the non-specialist who wishes to gain some deeper understanding of the issues. In the years leading up to our retirements, Robert and I ran and administered a theological training program at Bristol Baptist College in the UK to help ordinary Christians relate their faith to modern life, and this book seeks to achieve a similar purpose in the area of nuclear power. A few may find some chapters more detailed than they would like (while at the same time others will value the breadth of data that has been gathered here), but the summarizing conclusions will point out the essence of the discussion. There are also suggestions for further reflection, making this a useful tool for a group working on issues around energy policy and/or environmental concerns. If one or two churches also find the book valuable for engaging with ethical issues around the use of nuclear power, I am sure the author will feel he has achieved his aim.

Kenneth Stewart
Former Director of Pastoral and Mission Studies
Bristol Baptist College

Preface

My book is a non-specialist work for those who want to know more about nuclear energy. Specifically, I have written from a Christian perspective as it appears to me that there is a gap here for Christians who want to be more informed. Often, nuclear energy is regarded with foreboding, silence, and suspicion. For too long a negative nuclear narrative has prevailed in the debate/dialogue on the role of nuclear energy in the UK. I present a positive nuclear narrative that comes from my experience of working for many years as a scientist on nuclear plants and a tutor in a training center for the UK's nuclear industry. During this time, I have been a committed Christian and, in fact, completed my part-time theology degree while working as a tutor. When I left my full-time employment, it was to start postgraduate degrees in biblical studies at the University of Bristol, where I completed my PhD in New Testament. After these studies, I returned to the nuclear industry working as a consultant on various projects. At the same time, I did some teaching in New Testament and New Testament Greek. My final years before retirement were spent in the rich atmosphere at Bristol Baptist College as a course administrator and tutor in New Testament Greek (teaching Anglican and Baptist students). Putting all this together, I wanted to research and write in retirement on the benefits of nuclear energy. This book is the outcome. I include footnotes and a large bibliography, for it seems to me that on such a controversial topic readers should know my sources and follow them up if they wish. However, it is possible to read the book without reading most footnotes. I provide further reading and questions for either individual study or group discussions. And although

PREFACE

my focus is on the UK, I include other countries, particularly the United States, in my discussion to give a wider picture of the world situation. You will notice that I use *World Nuclear News* frequently as this provides regular updates in short articles. Most of my resources are available on the web. Enjoy your reading, thinking, and reflection as you are challenged on the merits of nuclear energy in a nuclear-powered universe which God created.

Acknowledgments

THIS BOOK COULD NOT have been written without the help of many people and organizations. Before my retirement I worked at Bristol Baptist College, in Bristol, England, as a course administrator and tutor in New Testament Greek. When the idea of writing a book on nuclear energy came to mind I received support from staff and students. Such an idea arose from my long experience of working in the UK's nuclear industry. In particular, I wish to thank my former colleague Revd Kenneth Stewart (then Director of Pastoral and Mission Studies), who encouraged me and read sample chapters for submission to prospective publishers. Since retirement he has commented on all my draft chapters, giving valuable feedback including exhorting me to write at the level for a non-specialist! This I have aimed to do. He has also written the foreword.

Further, thanks are due to Mike Brealey (Librarian) for providing access to the Baptist College library and support in locating books. Fran Brealey (College Manager) provided assistance on locating module information for the University of Durham Joint Awards programs which are currently studied by Baptist College students (and also Anglican students at Trinity College, Bristol). Revd Dr. Ernest Lucas (Vice-Principal Emeritus), whose teaching/writings on science and theology encouraged my thinking, suggested that I approach Wipf and Stock as publisher. I am grateful for his advice. Thanks also to Emeritus Professor of Biochemistry at Bristol University Andrew Halestrap for his interest in my project and suggesting endorsers. And, I appreciate the assistance of librarians in the public library system, at Bristol, Paignton, and Churston, for locating books I requested.

ACKNOWLEDGMENTS

Although it is some years since I worked in the nuclear industry as full-time staff, and then as a consultant, I retain professional membership of the Society for Radiological Protection and the Nuclear Institute. Their newsletters, journals (*Journal of Radiological Protection* and *Nuclear Future*), and meetings keep me informed. Further, as a member of the union *Prospect* I benefit from its focus on energy, including the nuclear industry. The Retired Professional Engineers Club (Bristol) has provided an interesting range of relevant talks on energy issues which I have attended. The University of Bristol, too, has provided valuable free public lectures on climate change and a two-day seminar on the Advanced Boiling Water Reactor and Underpinning Nuclear Energy (May 2016), which I attended. Moreover, I was privileged to participate in the UK's public consultation National Geological Screening, led by Radioactive Waste Management on the approach for a future geological disposal facility (GDF) for radioactive waste. National geological screening considered information on UK geology relevant to the long-term safety of a GDF.

Redcliffe College, Gloucester, and the John Ray Initiative, both Christian organizations, ran conferences on environment/energy which I attended. Moreover, church discussions on climate change, creation care, and feedback from the important COP21 contributed to my thinking. I am grateful to our men's evening Bible study group for support and prayer.

I am indebted for the material readily available on the Internet from individuals, churches, charities, organizations (national and international), governments, and regulators. Thanks to open access I read important articles which otherwise I would have missed. The UK's Open Government Licence is also appreciated. I read more than I cited, but it helped in my thinking.

Often people write about "the nuclear industry" in a way that some people write about things, but we must not forget that "the nuclear industry" is not just about technology; it is also about people, whom God loves. People in the industry move on, age, retire, and die. I therefore acknowledge the colleagues and friends I worked with in the industry over the years, in particular, the late John Edwards, who was my team leader when I was a young scientist in his health physics commissioning team at Hinkley Point B Power Station. I learned much from them about energy technology, safety, values, culture, and their families. And before them I appreciate my school and university teachers who developed my interest in science. Further, it would be remiss of me if I failed to acknowledge all my teachers who supported my theological education over the years, and especially the

ACKNOWLEDGMENTS

support given by the late Dr. Diane Treacy-Cole at the University of Bristol. who became my PhD adviser in New Testament.

Thanks to my proofreaders: Kenneth Stewart (see above); my friend Geoffrey Sutton (retired languages teacher), who ensured I addressed non-specialists, not science graduates or academics; and our son, Tim, who found typos, asked questions for clarification, and made suggestions. Geoffrey spent many hours improving drafts, identifying typos, asking good questions, and printing pages. Also, thanks go to their wives, who gave them the time for this.

Also, considerable gratitude goes to my wife, Susan, for her support over a lifetime of studying and then over many months of writing when we could have been busy spending time together. Our two daughters, Laura and Jess, with their husbands and grandchildren also understood when I was busy, as did our son, Tim. Hopefully, one day our grandchildren may read this book, appreciate my concerns for caring for creation, and enjoy—with many others—our shared world.

Many thanks to Wipf and Stock for offering to publish this book on a controversial but important area for debate and dialogue. They were willing to publish based on merit rather than projected sales. Thanks in particular go to Matthew Wimer, Assistant Managing Editor, who promptly answered my questions during the research and writing stages. When I applied for a two-month extension he quickly agreed to this without putting me under any pressure to finish in my original timescale. Moreover, because of the fast-changing nature of the nuclear industry, he kindly agreed to me updating my manuscript after I submitted it in October 2016. This enabled me to include important statements and sources into May 2017. Nathan Rhoads, my copy editor, worked in meticulous detail on my text with patience as he raised questions and I answered his various queries. He also took on my updates to the manuscript. Jana Wipf provided the typeset pages and included further updates. Shannon Carter designed the cover. Brian Palmer completed the contract requirements. I am grateful.

I am indebted to all these people who have helped bring this book to fruition. Every effort has been made to attribute sources in the bibliography, but please contact the publishers if you have any queries. Any errors of omission or commission remain my responsibility. Finally, thanks be to God for the opportunity to complete this writing in combining both issues of faith and science. As Professor of Physics at the UK's Durham University Tom McLeish sees it: science is God's gift to us (as stated in his presentation "Faith and Wisdom in Science (with Job)," delivered for Christians in Science Bristol, March 10, 2017).

Abbreviations

2DS	2 degrees C scenario
AAAS	American Association for the Advancement of Science
ABWR	Advanced Boiling Water Reactor
ACP100	a Chinese SMR design
ACSWP	Advisory Committee on Social Witness Policy (of the Presbyterian Church, USA)
Acts	Acts of the Apostles (a book in the New Testament)
AGR	advanced gas-cooled reactor
AIP	American Institute of Physics
A-Level	Advanced Level
ANS	American Nuclear Society
AP1000	Advanced Passive reactor, a PWR design
AR5	Fifth Assessment Report of the IPCC
BBC	British Broadcasting Corporation
BEIS	Department for Business, Energy and Industrial Strategy
BIS	Department for Business, Innovation and Skills
BMS	Baptist Missionary Society
Bq	becquerel (SI unit of radioactivity)
BWR	boiling water reactor
C-14	carbon-14
CANDU	Canada's Deuterium Uranium reactor design
CCC	Committee on Climate Change
CCS	carbon capture and storage

ABBREVIATIONS

Cefas	Centre for Environment, Fisheries and Aquaculture Science
CEGB	Central Electricity Generating Board
CEL	Christian Ecology Link
CERN	Conseil Européen pour la Recherche Nucléaire (European Council for Nuclear Research)
Ci	curie (earlier unit of radioactivity)
CiS	Christians in Science
CMI	Chartered Management Institute
CNC	Civil Nuclear Constabulary
CNS	Canadian Nuclear Society
CNSP	Civil Nuclear Security Programme
CO_2	carbon dioxide
COC	Committee on Carcinogenicity of Chemicals in Food, Consumer Products and the Environment
COP21	UNFCCC Conference of the Parties, 21st session, held in Paris, November 30 to December 11, 2015
CPNI	Centre for the Protection of National Infrastructure
CT	computed or computerized tomography
DAC	Design Acceptance Confirmation
DECC	Department of Energy and Climate Change
DOE	Department of Energy (US)
DoSER	Dialogue on Science, Ethics, and Religion (a program of the AAAS)
DTI	Department of Trade and Industry
EA	Environment Agency
EC	European Commission
EESC	European Economic and Social Committee
EIA	Energy Information Administration (of the US DOE)
EPA	Environmental Protection Agency (title used in different countries, e.g., Ireland)
EPR	European Pressurised Reactor, a PWR design
EPSRC	Engineering and Physical Sciences Research Council
ERDO	European Repository Development Organisation
ESC	Electrical Safety Council
et al.	*et alia*, Latin for "and others" (used in references to indicate that there are more authors than those named)
EU	European Union

ABBREVIATIONS

Euratom	European Atomic Energy Community
FOE	Friends of the Earth
FOE Cymru	Friends of the Earth Cymru (Wales)
FORATOM	European Atomic Forum
FSA	Food Standards Agency
GCSE	General Certificate of Secondary Education
GDA	Generic Design Assessment
Gen	Genesis (book in the Old Testament)
GHG	Greenhouse Gas
GWe	gigawatts of electricity (1,000,000,000 watts, or 1,000 megawatts)
GWPF	Global Warming Policy Foundation
Gt	gigatonnes (1,000,000,000 tonnes or 1,000 megatonnes, where 1 tonne = 1,000 kg). In the United States the term metric ton is used for tonne. The United Kingdom, and other countries, sometimes use the term ton (2,240 pounds). As the difference between tonne and ton is small I make no distinction between them.
H-3	tritium
HCBEISC	House of Commons Business, Energy and Industrial Strategy Committee
HCECCC	House of Commons Energy and Climate Change Committee
HCHLJC	House of Commons and House of Lords Joint Committee on the National Security Strategy
HCSTC	House of Commons Science and Technology Committee
HCWAC	House of Commons Welsh Affairs Committee
HEU	highly enriched uranium
HLSTC	House of Lords Science and Technology Committee
HMG	Her Majesty's Government (the government of the UK)
HPA	Health Protection Agency
HSE	Health and Safety Executive (in UK) or Health Service Executive (in Ireland)
I-131	iodine-131
IAEA	International Atomic Energy Agency

ABBREVIATIONS

Ibid.	*ibidem*, Latin for "in the same place" (used to reference the source given in the preceding citation)
ICRP	International Commission on Radiological Protection
IEA	International Energy Agency
IET	Institution of Engineering and Technology
ILW	intermediate-level waste
IMechE	Institution of Mechanical Engineers
INES	International Nuclear and Radiological Event Scale
INPO	Institute of Nuclear Power Operations
IPCC	Intergovernmental Panel on Climate Change
ITER	International Thermonuclear Experimental Reactor
JRI	John Ray Initiative
K-40	potassium-40
kg	kilogram (1,000 grams)
km	kilometer (1,000 meters)
LNT	linear no-threshold
m	meter
m^{-3} or $/m^3$	per cubic meter
Matt	Matthew (a book in the New Testament)
MIT	Massachusetts Institute of Technology (US)
MOX	mixed oxide fuel
mrem	millirem (earlier unit of radiation dose)
MSc	Master of Science degree
Mt U	megatonnes of uranium (1,000,000 tonnes, where 1 tonne = 1,000 kg)
MWe	megawatts of electricity (1,000,000 watts)
mSv	millisievert (1/1,000 sievert)
μSv	microsievert (1/1,000,000 sievert)
N-14	nitrogen-14
NAE	National Association of Evangelicals (US)
NASA	National Aeronautics and Space Administration (US)
NCfN	National College for Nuclear
NCRP	National Council on Radiation Protection and Measurements (US)
NCSL	National Conference of State Legislatures
NDA	Nuclear Decommissioning Authority
NEA	Nuclear Energy Agency (of the OECD)
NEI	Nuclear Energy Institute (US)

ABBREVIATIONS

NFCRC	Nuclear Fuel Cycle Royal Commission (South Australia)
NGL DNB PLEX	EDF Energy Nuclear Generation Ltd (NGL) Dungeness B (DNB) Plant Life Extension (PLEX)
NHS	National Health Service
NI	Nuclear Institute
NIA	Nuclear Industry Association
NIC	Nuclear Industry Council
NIRAB	Nuclear Innovation and Research Advisory Board
NNL	National Nuclear Laboratory
NNSA	National Nuclear Security Administration (of the US DOE)
NORM	naturally occurring radioactive material
NRC	Nuclear Regulatory Commission (US)
NRW	Natural Resources Wales
NSAN	National Skills Academy for Nuclear
NSS	National Security Strategy
NSSG	Nuclear Skills Strategy Group
OECD	Organisation for Economic Co-operation and Development
ONR	Office for Nuclear Regulation (UK)
OTC	once-through fuel cycle
OU	The Open University
Pb-210	lead-210
PHE	Public Health England
Po-210	polonium-210
PSR	Periodic Safety Review
PWR	pressurized water reactor
R&D	research and development
Ra-220	radon-220, also called thoron
Ra-222	radon-222
RAE	Royal Academy of Engineering
RPII	Radiological Protection Institute of Ireland
RWM	Radioactive Waste Management
SDGs	Sustainable Development Goals (a UN initiaive)
SI	Système International d'Unités (International System of Units)
SMC	Science Media Centre

ABBREVIATIONS

SMR	small modular reactor
SNP	Scottish National Party
SoDA	Statement of Design Acceptability
SPRU	Science and Technology Policy Research (University of Sussex)
SRP	Society for Radiological Protection
SRTP	Society, Religion and Technology Project (Church of Scotland)
STEM	science, technology, engineering, and mathematics
Sv	sievert (SI unit of ionizing radiation dose)
TENORM	technologically enhanced (industrial) naturally occurring radioactive material
Th-232	thorium-232
TSO	The Stationery Office
TTC	twice-through fuel cycle
tU	tonnes of uranium (where 1 tonne = 1,000 kg)
TUC	Trades Union Congress
TVA	Tennessee Valley Authority (US)
TW	terawatt (1,000,000 megawatts)
TWh	terawatt hour
U-234	uranium-234
U-235	uranium-235
U-238	uranium-238
UK	United Kingdom
UK EPR	UK European Pressurised Reactor, which EDF Energy is building at Hinkley Point C
UN	United Nations
UNFCCC	United Nations Framework Convention on Climate Change
UNSCEAR	United Nations Scientific Committee on the Effects of Atomic Radiation
US	United States of America
USNIC	United States Nuclear Infrastructure Council
UV	ultraviolet light
WANO	World Association of Nuclear Operators
WARP	Where Are the Radiation Professionals? (an NCRP initiative)
WCC	World Council of Churches

ABBREVIATIONS

WEA	World Evangelical Alliance
WEC	World Energy Council
WEF	World Economic Forum
WENRA	Western European Nuclear Regulators' Association
WHO	World Health Organization
WiN UK	Women in Nuclear UK (branch of the NI)
WIPP	Waste Isolation Pilot Plant
WNA	World Nuclear Association
WNN	*World Nuclear News*
YGN	Young Generation Network

Introduction

> Given all that we know, it is therefore very evident to many Christians that the need to address climate change and its effects on human life and environment must be one of the central aspects of our Christian calling today. And many have taken this seriously.
>
> ELAINE STORKEY[1]

Electricity, Energy, and Environment

ELECTRICITY POWERS THE TECHNOLOGY in our homes when we use our computers, electric lights, freezers, refrigerators, telephones, televisions, and washing machines. Where does it come from? And the same question arises for technology outside our homes (e.g., street lighting, traffic lights, industry). Is our electricity supply secure, sustainable, and affordable? We take electricity for granted and generate it. Does it contribute to climate change and global warming? All people worldwide (and non-human species) are potentially impacted by adverse changes to our environment, and decisions we make on future energy policies are crucial factors.

In a democracy we have the opportunity of contributing to this decisive debate and dialogue, but we need to have a firm foundation on which to make sensible decisions. We need to recognize the benefits and

---

1. Storkey, "Environment and the Developing World," 127. Dr. Elaine Storkey was President of Tearfund, the UK Christian relief and development agency, from 1997 to 2013. She is a theologian, author, speaker, and broadcaster. See http://www.elainestorkey.com.

risks inherent in various technologies for generating electricity, but also unfortunately producing pollution. To be sure, electricity production is not the only contributor to carbon dioxide emissions and other global warming gases. But, important as these areas are (e.g., transport emissions), my scope is limited to comments on decarbonization—my focus is on nuclear energy as part and parcel of electricity production.

Nuclear is in the news—for good or ill. Nuclear energy is controversial, raising key questions, such as: Is nuclear energy for generating electricity environmentally sound, economic, efficient, and ethical? Is nuclear energy evil, even demonic? Are nuclear energy's advocates liars, deceivers, misled by technology, or are they good guys concerned about the Earth? These are some of the issues which we consider. Nuclear accidents such as the one at Fukushima (Japan) naturally raise fears, but they should not close down the debate. All technologies employed for generating electricity have their pros and cons, and these should be addressed in a proportionate manner. In this book I put forward a Christian case for supporting civil nuclear power plants as part of an energy mix in the United Kingdom (UK) and other countries.[2] I don't suggest that all of our electricity should be generated by nuclear energy or that none of our electricity should be nuclear generated. Rather, my approach is to argue for nuclear as a vital part of the UK energy policy. However, this topic is not just about technology but concerns people, populations (human and non-human), and our planet in the short and long term.

In the UK nuclear energy is part of school education (e.g., science and religious education) on energy sources and energy policies in both GCSE and Advanced Level (A-Level) subjects (e.g., physics).[3] Moreover, university undergraduates and postgraduates are interested in engaging with these issues and some courses/modules particularly address them in depth. However, the wider population, including Christians, is aware of nuclear energy issues, debates, and dialogue relevant to not only the UK, but to Europe and worldwide.

Many authors have written on the potential impact of climate change on our planet and all its creatures.[4] Christians and theologians have written

2. By "civil" (or commercial) nuclear power plants I mean the non-military uses of nuclear plants to produce electricity. Military submarines that use nuclear reactors are outside the scope of this study.

3. GCSE (General Certificate in Secondary Education) courses have examinations at sixteen years of age. A-Level subjects have examinations at eighteen years of age. On books, for example, see Hulme and Taunton, *GCSE Physics: Revision Guide*, 27–36 and 133–48.

4. See, for example: Dyer, *Climate Wars* (2010); Lynas, *Six Degrees* (2008); and Marshall,

widely, and positively, on creation care but there is less available specifically on nuclear energy as an important, and positive, energy source.[5] My study aims to rectify this situation by helping to fill this gap. It is written from my Christian perspective as a scientist. I previously worked within the nuclear power industry at two nuclear power plants and then as a tutor at a training center. Moreover, I hold degrees in theology, biblical studies, and mission.

Christians do have a range of views, which the Church of Scotland's Society, Religion and Technology Project (SRTP) clearly captured in its publication *Energy for a Changing Climate*. It stated: "The Church of Scotland has thus far left open the question of nuclear power, aware of the differing views among church members, some strongly in favour, some strongly against, many undecided."[6] It then continued to discuss nuclear power and in 2010 produced its report "What Future for Nuclear Power?"[7]

Some authors admit to being reluctant converts to nuclear energy. For example, the British sociologist Anthony Giddens admitted: "Nuclear energy is mired in controversy, but . . . it is difficult to see how it will not figure in a prominent way—not for all industrial countries, but certainly for some of them."[8] Moreover, for Britain: "Since the proportion of electricity generated from renewable sources is so small, it is difficult to see how the UK could possibly meet its EU 2020 target of 16 percent from renewables if nuclear were allowed to lapse." Giddens doesn't want to downplay risks, but writes: "like many others, I am a reluctant convert to nuclear power, at least in so far as some of the industrial and developing countries are concerned."[9] He recognizes that there is currently no substitute for nuclear energy; removing nuclear from the energy mix is too great a risk. These are important issues that my book addresses.

As this is an introductory text for non-specialists, I avoid over-complicated scientific and technical discussions and mathematics, but provide further reading for interested readers. However, it addresses a controversial topic: civil nuclear energy. This topic is not an isolated subject since it

---

Don't Even Think About It (2015).

5. For example: Bookless, *Planetwise*; Houghton, *Global Warming*; and Spencer and White, *Christianity, Climate Change*. The John Ray Initiative has briefing papers available as downloads (www.jri.org.uk/briefings/).

6. Church of Scotland, *Energy for a Changing Climate*, 15, para. 5.6.2.

7. SRTP, "What Future for Nuclear Power?" Addressed later in the book.

8. Giddens, *Politics of Climate Change*, 132.

9. Ibid., 133.

engages science, technology, ethics, economics, environment, politics, and Christian mission.

My position is that nuclear energy should contribute to the UK's energy mix. Resisting the building of new nuclear power plants and refusing to keep the nuclear option open seems shortsighted. Nuclear energy has a positive contribution to make to electricity production in limiting climate change, where a growing global population, increased energy demand, and expectations are putting considerable strain on our planet and living species. Our children and future generations should not be disadvantaged because misinformation and misguided fears have prejudiced our debates, dialogues, and decision-making on nuclear energy. These issues are addressed in the context of the safety of various sources of electricity generation.[10]

I bring my expertise to give a Christian case that is justifiable and, although I am not impartial, I present my case in a fair and balanced way, respecting views different than mine.

Conclusion

So, can a viable Christian case be constructed for supporting the role of nuclear energy in the UK's energy mix? This introductory book aims to present the case for positive Christian engagement. The climate crisis is imminent, so we cannot just "wait and see" what happens. The UK and the world need long-term energy strategies to address the climate change crisis. Sir John Houghton clearly stated four reasons why such action is needed: scientific, economic, political, and spiritual.[11] He responsibly concluded: "Environmental justice is a spiritual discipline of faithfulness that comes from knowledge of the facts and a response of the heart."[12] So let us turn to chapter 1 and look at "Our Nuclear and Radioactive Universe."

Further Reading

Prance, Ghillean. *Thinking About . . . Creation Care: Is It the Responsibility of Christians to Care for the Environment?* Thinking About . . . leaflet series. Christians in Science, 2016. http://www.cis.org.uk/resources/thinking/.

10. MacKay remarked: "Even if we have no guarantee against nuclear accidents in the future, I think the right way to assess nuclear is to compare it objectively with other sources of power." MacKay, *Sustainable Energy*, 168.

11. Houghton, "Changing Global Climate," 213–15.

12. Ibid., 215.

1

Our Nuclear and Radioactive Universe

How old are you? Seventeen? Eighteen? No, billions of years old. We are all made from atoms and, apart from hydrogen (there is no helium in our bodies), the nuclei of these atoms were made in nuclear fusion reactions inside stars that exploded billions of years ago. We are, literally, 'stardust'.

STEVE ADAMS AND JONATHAN ALLDAY[1]

Then the Lord God formed a man from the dust of the ground and breathed into his nostrils the breath of life, and the man became a living being.

GEN 2:7

By the sweat of your brow you will eat your food until you return to the ground, since from it you were taken; for dust you are and to dust you will return.

GEN 3:19

---

1. Adams and Allday, *Advanced Physics*, 528. One billion years is one thousand million or 1,000,000,000 years. Nuclear fusion is the process by which the nuclei of light elements (such as hydrogen or helium) are forced together and they form a heavier nucleus. In the process, a vast amount of energy is released. See Prescott, *Oxford Study Science Dictionary*, 171. Fusion is covered in school science topics on understanding the stars.

Let There Be Light!

Introduction

IN THE INTRODUCTION, WE identified the importance of understanding nuclear energy's vital role in a balanced energy mix from the perspective of presenting a Christian case for embracing nuclear energy positively. I mentioned the good work that has been done by Christians, for example on creation care and climate change, but I identified a gap in specific discussions of nuclear energy as an important, and positive, energy source.

This chapter considers our universe as the correct context in which to address nuclear energy from a Christian perspective. Christians believe, because Christian theology teaches (as revealed in the Scriptures), that God is the creator of our universe. This chapter introduces a basic scientific understanding of our universe with a focus on its necessary nuclear resources and natural radioactivity.[2]

Nuclear energy and radioactivity, as we shall see, are part and parcel of our created universe. Since nuclear energy and radioactivity from nuclear power stations are frequently viewed with suspicion, we should first look wider to understand the essential structure of the universe from the past to the present. How does this understanding relate to nuclear energy? First, let us turn to the book of Genesis.

Genesis 1:1 starts: "In the beginning God created the heavens and the earth," and 1:3–4 states: "And God said, "Let there be light," and there was light. God saw that the light was good, and he separated the light from the darkness." This was the first day. A little later, we read that on the fourth day God put lights in the sky:

> God made two great lights–the greater light to govern the day and the lesser light to govern the night. He also made the stars. God set them in the vault of the sky to give light on the earth, to govern the day and the night, and to separate light from darkness. And God saw that it was good. (1:16–18)

On the sixth day, "God saw all that he had made, and it was very good" (1:31).

Christians understand that God is the creator. Reading Genesis in its cultural setting, Ernest Lucas asks why—when there were specific words in

---

2. Some readers will be aware of different definitions for the universe. For example, the physicist and cosmologist Paul Davies, *Goldilocks Enigma*, 35–38, discusses five definitions: the observed universe, the observable universe, the entire universe, the pocket universe, and the multiverse. Discussing these is beyond the scope of my study. See Weaver, *Christianity and Science*, 69–71.

Hebrew for them—the sun and moon are only referred to as "lights." He stated:

> The probable answer is that, in the Semitic languages, of which Hebrew is one, the words "sun" and "moon" are also the names of gods. The peoples around the Hebrews worshipped the heavenly bodies as gods and goddesses. Genesis 1:14–19 is an attack on all such thinking.[3]

Genesis rejects polytheism. The sun and moon are "lights" created by God; they are not deities to be worshipped. It is a theological statement. They are part of God's material creation, which is pronounced "good" and "very good."[4] And these statements appear before what Christians call the "fall" (humankind's disobedience to God) in Genesis 3. The issue of the fall, some Christians believe, caused bad changes and the restructuring of nature. However, "nuclear" is not a punishment because of humankind's sin. It was God's plan in creation to use nuclear energy. I will show that nuclear energy and radioactivity are a normal part of God's creation in our universe.

Some readers may want to ask about the relationship between science and religion. Although this interesting debate is beyond the scope of my study, I encourage interested readers to read the work of scientists who are also Christian scholars (see Further Reading at the end of this chapter).

This chapter considers the Big Bang, stars, nuclear fusion, nuclear fission, and natural radioactivity. We will see that God uses nuclear fusion to power his universe giving us heat and light. The stars are the origin of the elements from which we are created. Our universe is naturally radioactive. The natural nuclear reactors which generated nuclear energy from fission began about two billion years ago in the uranium mines at Oklo, West Africa. Finally, I introduce natural radioactivity which is present on Earth. In this radioactive environment human beings and non-human species live. Scientists understand that in this nuclear-powered and radioactive

---

3. Lucas, *Can We Believe Genesis Today?*, 99. Further, see his *Interpreting Genesis*, 3. Also, see Weaver, *Christianity and Science*, 43–44. McKeown, *Genesis*, 24, agrees with Lucas.

4. Walton, *Lost World of Genesis One*, 152, sees God as creator, but indicates: "Genesis 1 presents an account of functional origins and . . . it offers no descriptive mechanism for material origins." Moreover, on the refrain "it was good," Walton refers to "the functional readiness of the cosmos for human beings. Readers were assured that all functions were operating well and in accord with God's purposes and direction" (50). Space only permits me to briefly consider the Genesis creation narrative, but it is worth mentioning that in the Bible there are other creation stories, e.g., in Proverbs 8 and Psalms 19, 33, and 104.

universe, life and human beings live and evolve. Christian scholars too support this understanding. Clearly, my focus is specifically on nuclear energy and there is no intention to describe aspects of the universe in the depth discussed by astronomers (fascinating though this is). My intention is much more modest. It is a simple introduction relevant to my particular focus. Let us now begin with the Big Bang.

From the Big Bang to Our Solar System: Origin and Structure of the Universe

Professor Brian Cox and Andrew Cohen, in their volume to accompany the BBC's spectacular TV series *Wonders of the Universe*, commented: "At 13.7 billion years old, 45 billion light years across and filled with 100 billion galaxies—each containing hundreds of billions of stars—the universe as revealed by modern science is humbling in scale and dazzling in beauty."[5] Nuclear-powered stars create the chemical elements that form us human beings; as Steve Adams and Jonathan Allday stated in this chapter's opening quotation, we are "stardust." Cox and Cohen explain this process of nuclear-powered stars (see below).

After the Big Bang, within fractions of a second, there existed energy, particles, and immense heat. Space expanded and the temperature began to drop. Within about a millionth of a second, protons and neutrons were formed, and then after three minutes helium nuclei were formed. Electrons existed, but atoms weren't formed because energy levels and temperatures were too high. However, after about 400,000 years, when the temperature had cooled to 2,700°C, electrons began orbiting a nucleus, and atoms were formed.

The simplest hydrogen atom has one proton in its nucleus and it is orbited by one electron. Heavy hydrogen (deuterium) has both a proton and a neutron in its nucleus.[6] Helium-4 atoms have two protons and two neutrons in their nucleus. This nucleus is surrounded by two orbital electrons (much like planets orbit the sun). At this stage, most atoms were hydrogen

---

5. Cox and Cohen, *Wonders of the Universe*, 8.

6. You may recall from school science that a proton is positively charged and has almost the same mass as the neutron, which has no electrical charge. The electron is negatively charged and has a much smaller mass than the proton or neutron. Electrons orbit the central nucleus of protons and neutrons (except for hydrogen-1, which has only one proton in the nucleus and no neutrons).

and about 25 percent were helium. Clouds of gas, acting under gravity, created the first stars (formed many millions of years after the Big Bang).[7] Light started shining from the energy released in nuclear fusion. Eventually galaxies were formed. Within the Milky Way galaxy, our solar system was formed about 4.56 billion years ago.

This generally accepted scientific viewpoint is also accepted by the scientist and theologian John Weaver in *Christianity and Science* (see Further Reading). Paul Davies, physicist and cosmologist, explains this process in some detail in his book *The Goldilocks Enigma: Why Is the Universe Just Right for Life?*[8]

The Sun

Where does the sun's energy come from? Back in the nineteenth century one scientist, William Thompson—later called Lord Kelvin—attempted to date the Earth by working out the sun's heat source. Kelvin, a Christian, asked why over the life of the Earth (presumed as thousands of years) the sun had not cooled down. Colin Russell remarked: "The ordinary laws of heat suggest a glowing solid sphere should have cooled down long before now. No chemical reactions were known generating such energy, and nuclear energy had yet to be discovered."[9] Kelvin included other factors to try and account for this problem.

Today, we know that nuclear fusion powers the sun, which is mainly hydrogen and helium (70 percent and 28 percent, respectively, with just 2 percent other elements). It is not powered by chemical reactions or nuclear fission. Most of the power generated comes from the fusion of four hydrogen nuclei into one helium-4 nucleus and this releases a vast amount of energy. Actually, this occurs in three steps but the overall reaction in the core of the sun is four protons fusing to produce one helium nucleus (consisting of two protons and two neutrons). This produces a tremendous amount of heat, generating an outward pressure which counteracts the force of gravity acting inwards. Cox and Cohen observed that every second

---

7. See, in Cox and Cohen, *Wonders of the Universe*, the helpful "Timeline of the Universe: The Big Bang to the Present" (110–11) and "Life Cycle of the Universe" (222–23). Davies, *Goldilocks Enigma*, 152, mentions the first stars started forming after "several hundred million years."

8. See ch. 2, "The Universe Explained," and ch. 3, "How the Universe Began."

9. Russell, *Earth, Humanity, and God* (1994), 32.

our sun "converts 600 million tonnes of hydrogen into helium."[10] And this amount of energy released in one second from nuclear fusion is more energy than humankind will consume for the next million years. The fusion energy in the core eventually results in emissions from the sun's surface of visible light, infrared and ultraviolet radiation, x-rays, etc. The sun's light is reflected by our moon, which then illuminates the Earth. The moon has no energy source itself to produce light.

Scientists reckon that the sun will continue this fusion process for about another five billion years, producing life-giving heat and light on Earth. Once the hydrogen is used up in nuclear fusion, then the nuclear reaction will continue as the remaining helium undergoes nuclear fusion. As the sun's hydrogen store is consumed the sun will expand and become a red giant. After its helium shell is consumed the sun will collapse and become a white dwarf. So, our sun and other stars have a life cycle from birth to death. The nuclear reaction begins at the point of ignition but eventually our sun, after about ten billion years, will use up its nuclear fuel.

Stars and Supernovae: Origin of the Elements

The sun, powered by nuclear fusion, is our nearest star at 93 million miles (150 million km) from Earth. Nuclear fusion also powers all the other stars, although not all stars are identical. They vary significantly in size, mass (the amount of material in them), color, brightness, and lifespan. However, there is a pattern to the stars. In the early twentieth century two astronomers (Hertzsprung and Russell), working separately, plotted a graph of the surface temperature of stars against their brightness and noticed a pattern. Most stars lie on what is called the Main Sequence of the graph. Our sun lies somewhere around the middle. What processes power the stars? Well, the physics is straightforward. These Main Sequence stars, and other stars, are all powered by nuclear fusion reactions. The stars contain the gases hydrogen and helium, which undergo fusion.[11]

---

10. Cox and Cohen, *Wonders of the Universe*, 117.

11. This famous graph is called the Hertzsprung-Russell diagram. Most stars lie on what is called the Main Sequence, which is a line ascending from the bottom right of the graph (where the surface temperatures and brightness are lowest) to the top left, where the surface temperatures are higher and the brightness is greatest. For the Hertzsprung-Russell diagram with a good explanation, see Cox and Cohen, *Human Universe*, 97–99. At the time of writing *Human Universe* (2014), Professor Brian Cox was at the University of Manchester and was a researcher at the Large Hadron Collider (Switzerland). Andrew Cohen was

This means that when you look out at the stars at night you see the light from nuclear reactions. Even if some of these stars died millions or billions of years ago, their light is still reaching Earth. Again, Cox and Cohen inform us: "Fusion is the process that powers every star in the heavens, including our sun. Look up into a clear blue sky and you are bathing in the energy of nuclear explosions on an unimaginable scale."[12] Our universe is truly powered by nuclear energy.

But nuclear fusion doesn't just give us light and heat; it does far more! Immediately after the Big Bang only simple atoms such as hydrogen and helium were formed, so where did all the other elements we now know (such as carbon, oxygen, and iron) come from? If we just consider carbon atoms, then each person contains billions upon billions of them. What is their origin? They arose through a process called nucleosynthesis, which simply means that nuclear fusion reactions in the stars are the origin of the chemical elements shown in the periodic table.

Stars have major phases from their birth to their death, but it is beyond the scope of my study to discuss this in detail. Suffice it to say that once a star has consumed most of its hydrogen then its core is mainly helium. The force of gravity collapses the star's core and its temperature rises so that it is able to fuse helium. The energy generated stops the star's collapse; it stabilizes but also swells in size. The star becomes a red giant and the process leads to the origin of the elements which will be spread to form nebula (clouds of gas and dust).

Fusing two helium nuclei (each made up of two protons and two neutrons) produces beryllium-8 (four protons and four neutrons), which in turn can fuse with another helium nucleus to give carbon-12 (six protons and six neutrons). This process gives us our carbon atoms. Further fusion of carbon and helium nuclei produces oxygen-16, essential for life on Earth. Eventually, all the helium in the star is consumed in the nuclear fusion processes. Then our sun, and other stars of similar mass, can no longer sustain the fusion process since the star's gravitational force is too small to compress the core and restart the nuclear fusion. The sun will eventually become a white dwarf after consuming all its helium.

However, massive stars have the capability through gravity to continue the fusion process and produce more heavy elements. Twenty-five elements are created in their core, but when iron (the twenty-sixth element)

---

head of the BBC Science Unit and an honorary lecturer at the University of Manchester.

12. Cox and Cohen, *Wonders of the Universe*, 116.

is reached the fusion process stops and a core of iron is formed. After iron, energy considerations require a vast amount of energy input, whereas previously in this fusion process energy has been released. Now the star collapses and these twenty-six elements are spread into space as debris, which we know as planetary nebulae.

But what about the other heavier elements beyond iron, such as copper, lead, gold, and uranium? Where do these originate? Once again the answer is nuclear fusion. Stars with a high mass, "dying giants," experience a catastrophic collapse of the iron core in a short timescale, which produces an extreme temperature rise with a tiny, very dense core. Protons and electrons combine to give a compressed ball of neutrons. The star experiences such severe conditions that the heavier elements (gold up to plutonium) are formed as the star explodes with an intense light. This is a supernova. The created heavy elements are spread out into space from the explosion. Such supernovae are rare events, but from the debris of dying stars new stars and planets are born. This cycle works through nuclear fusion and continues today. From these elements chemical compounds were formed in space so that in the Orion Nebula (on the edge of Orion's Belt), for example, we have complex molecules.

Some of the elements produced in stars and supernovae are radioactive. This means these elements are unstable and will emit particles and energy to become stable. For example, with another type of supernova, carbon and oxygen atoms can be fused to produce the radioactive element nickel-56 with a half-life of 60 days (the half-life is the time it takes the radioactive material to decay to half of its original activity). This radioactive decay makes them identifiable as their brightness decreases following this decay.[13] The metal uranium which is formed in supernovae and used as fuel in nuclear reactors is naturally radioactive. Radioactivity, like nuclear fusion, is a normal part of God's creation in our universe.

In summary, from such a nebula "five billion years ago, our sun was formed. Around that star a network of planets condensed from the ashes, and amongst them was Earth; a planet whose ingredients originated from the nebula, a cloud of elements formed in the deaths of stars, drifting

---

13. Adams and Allday, *Advanced Physics*, 541. Nickel-56 undergoes radioactive decay to cobalt-56 and then to iron-56. They discuss the two types of supernova, but we don't need to discuss this further. For an explanation of supernovae with a diagram, see Cox and Cohen, *Wonders of the Universe*, 128–33.

through space."[14] All this derives from nuclear fusion in the stars. We, and our world, are composed of stardust as part of God's original plan.

The Earth

Our Earth revolves (with other planets, etc.) around the sun in our solar system within the Milky Way galaxy. Scientists date the origin of the solar system and Earth's age at about 4.56 billion years using various recognized techniques, such as radioactive decay (radiometric decay).[15]

The Earth is a fragile planet facing many perils. In 1994 Colin Russell identified these perils as: flooding, volcanoes, earthquakes, comets, meteorites, radiation, and solar warming. He rightly asked what technological and theological responses were appropriate.[16]

The Earth is exposed to radiation from the sun and space (cosmic radiation), so life needs protection from damaging exposure. In what ways and from what? First, there are the Earth's natural defense mechanisms (its magnetic field, atmosphere, and ozone layer). Second, human beings have created protective procedures as we become more aware, through science, of the dangers in being exposed to the sun's rays. Each is considered below.

First, the Earth has its own protective mechanisms. There is its magnetic field, extending into space, which forms a blanket that helps protect us from charged particles coming from the sun (the solar wind, mainly composed of protons and electrons). Then we have the depth of our atmosphere (principally nitrogen and oxygen gas), which provides further shielding. The ozone layer in the upper atmosphere also filters out harmful ultraviolet radiation. If we travel in an airplane, then we reduce the effective protection of the atmosphere and expose ourselves to an enhanced amount of radiation. This applies to both aircrew and passengers.

Moreover, neutrons from cosmic radiation bombard our upper atmosphere and interact with nitrogen atoms to produce the radioactive element carbon-14, with a half-life of 5,730 years. This natural radioactive carbon-14

---

14. Cox and Cohen, *Wonders of the Universe*, 133.

15. A useful Christian and scientific discussion on the Earth's age is given by Robert White, *Age of the Earth*. He is Professor of Geophysics in the Department of Earth Sciences at Cambridge University. Also, Weaver, *Christianity and Science*, 100–101, identified radioactive elements used in radiometric dating (i.e., the dating of geological time).

16. Russell, *Earth, Humanity, and God*, ch. 4. Colin Russell was Emeritus Professor of History of Science and Technology at The Open University, Milton Keynes, UK.

diffuses into the lower atmosphere, where it enters living things. Some other radioactive elements are also naturally produced by cosmic radiation in the atmosphere, but carbon-14 is the most important. Carbon-14 is used by scientists to calculate the age of items.

If we leave Earth, as astronauts do, then we remove the protection provided by Earth and are exposed to higher radiation levels. This is a particular hazard for travel to, say, the moon or Mars. For now, my purpose is to identify that the Earth and our universe are naturally radioactive.

Second, we know that if we stare at the sun for long periods of time without protecting our eyes then the intense light will damage our eyesight and can lead to blindness. Also, we know it is painful to keep looking! We warn children not to stare at the sun. Yet even if we (sensibly) do not stare directly at the sun (remember the warnings that are given when an eclipse is being viewed not to look directly at the sun?) there are three other risks we are aware of: sunburn (erythema), skin cancer, and cataract formation. Substantial numbers of people are harmed by exposure.

Medical practitioners warn us not to expose ourselves to too much sunlight naturally (or via sun beds) since this exposure damages the skin and increase our risk of skin cancer. We can help protect ourselves by correctly applying and using the appropriate factor sun cream, limiting our time in the sun, and shielding our skin (e.g., arms) with clothing. Of course, exposure of our skin to the sun's rays will also cause the natural production of vitamin D, which is good for us—so we need a balance. We are encouraged to reduce the risk of cataract formation (opaque areas in the lens of the eye) by wearing suitable sunglasses to shield us from the invisible but damaging ultraviolet (UV) light. Suitable head covering also reduces the light intake to our eyes and gives further protection.

Living on Earth exposes us to these hazards from radiation and radioactivity. Moreover, as Colin Russell identified, the Earth and its inhabitants are vulnerable to a range of perils: flooding, volcanoes, earthquakes, comets, meteorites, radiation, and solar heating. The universe is a violent place. However, our focus here is limited to the naturally occurring radiation and radioactivity related to nuclear fusion. But there is another factor to consider: nuclear fission. Is this natural or not?

Nuclear Fission and Natural Nuclear Reactors

For the purpose of clarity, I wish to help readers understand the difference between nuclear *fusion* and nuclear *fission*. Both terms refer to what can happen within the nucleus of atoms. However, nuclear *fusion* occurs when lighter nuclei are joined together to produce a heavier nucleus (e.g., hydrogen atoms are forced, at very high temperatures, to fuse and make helium). In nuclear *fission* the opposite process occurs. Here a heavier nucleus is broken apart to give lighter nuclei (for example, uranium-235 is split apart, giving two lighter nuclei and releasing energy and neutrons). Nuclear fusion occurs in the stars. Nuclear fission occurs in our nuclear reactors.[17]

Nuclear fission, like *nuclear fusion,* is a *naturally occurring process.* Heavy nuclei (e.g., uranium and thorium) experience a low rate of what is termed spontaneous fission. Spontaneous fission occurs when a nucleus splits without any particle bombarding it—it just breaks apart by itself. In contrast, induced fission occurs when a neutron hits a nucleus and splits it apart, e.g., in uranium-235 within a nuclear reactor. Since neutrons occur naturally on Earth, then this natural fission process can occur without the modern technology of a nuclear reactor being built. Neutrons released in the fission process can go on and cause other nuclei to undergo fission, and a chain reaction is possible under certain conditions. Some neutrons will escape without causing further fission and these neutrons may be captured, making other elements become radioactive. For example, in nuclear power stations a neutron can be captured in a cobalt-59 nucleus (which is stable and not radioactive) to produce cobalt-60, which is radioactive, with a half-life of about five years.

Natural nuclear reactors were discovered in the seventies at Oklo (West Africa) in the uranium mines (located in Gabon). These reactors generated nuclear energy from fission about two billion years ago—a long time before people walked the Earth, let alone generated electricity from building nuclear reactors! And these natural reactors operated on and off over a period of hundreds of thousands of years. Scientists have investigated the sites and have discovered some remarkable facts about this natural process. The natural reactors operated because conditions were just right for fission chain reactions to operate. Three factors were favorable and necessary: (1) natural uranium ore with a high concentration of uranium-235,

---

17. For further information, see Prescott, *Oxford Study Science Dictionary,* "nuclear fission and fusion," 170–71. The dictionary is aimed at students in the 11–16 age group studying sciences.

(2) the presence of large quantities of water, and (3) the virtual absence of materials that captured neutrons (called neutron poisons) and prevented the chain reactions from continuing.

The first factor above (the uranium-235 concentration) needs some explanation. Modern reactors normally use "enriched uranium" fuel to sustain a chain reaction within the reactor core, since today the uranium ore we mine is composed of mainly uranium-238 (U-238), a smaller amount uranium-235 (U-235), and an even smaller amount of uranim-234 (U-234). Uranium-235 undergoes nuclear fission at a much higher rate than uranium-238 and so more uranium-235 is required for the chain reaction to be sustained. The amount of uranium-235 is increased by an enrichment process. However, uranium-235 is naturally radioactive, with a long half-life, whereas uranium-238 has a much longer half-life than uranium-235. So, uranium-235 decreases with time much more quickly than uranium-238 does (this decrease scientists traditionally call radioactive decay). This means that over the Earth's lifetime the ratio of these two uranium isotopes has changed: there is less uranium-235 in uranium ore now than previously (see Table 1). Water acts as a moderator. It slows down the fast neutrons that are given off in fission. Then these slow neutrons cause fission in U-235 and produce a chain reaction.

Thus, scientists have calculated that almost two billion years ago, when these natural fission reactors in West Africa operated, the percentage composition of uranium-238 and uranium-235 in uranium was different to the figures we have today. The uranium-238 percentage was lower while the percentage of uranium-235 was higher. This meant that the nuclear chain reaction could be sustained naturally.

So, the scientific evidence clearly shows that natural nuclear fission reactors are part of Earth's history long before human beings walked on our planet.[18] It just took us a while to find the evidence for the process.

---

18. See Davis et al., "Oklo Reactors." The authors provide a detailed scientific discussion. For more general reading see ANS, "Oklo's Natural Fission Reactors."

Uranium isotopes*	Half-life (years)	Percentage of uranium isotopes in uranium ore about two billion years ago	Percentage of uranium isotopes in uranium ore now
U-238	4.51 billion	Lower than today	99.27
U-235	710 million	About 4.0	0.72

Table 1: Uranium Isotopes: Half-Lives and Percentage Levels in Uranium Ore[19]

* Isotopes are the same chemical elements which exist with different numbers of neutrons in the nucleus. Here the chemical element uranium (U) has 92 protons in its nucleus and 92 orbital electrons. However, U-235 has 92 protons and 143 neutrons, whereas U-238 has 92 protons and 146 neutrons, i.e., three more neutrons in its nucleus than U-235 has in its nucleus. This difference gives U-235 and U-238 different nuclear properties, although they are both uranium and chemically the same.

Natural Radioactivity

As mentioned above when we considered the Earth, radioactivity occurs naturally. For example, carbon-14 is formed in our upper atmosphere by cosmic rays (neutrons) bombarding nitrogen gas (N-14) and releasing a proton. The nitrogen-14 is turned into radioactive carbon-14. Further, the Earth's rocks and soils are naturally radioactive from elements such as uranium and thorium. Some materials can cause significant radiation exposure to people since they contain higher levels of natural activity. These materials are called naturally occurring radioactive material (NORM). The Earth's oceans are also naturally radioactive. These sources of radiation exposure are called background radiation, and we discuss them further in chapter 2 together with artificial sources of radiation.

Living in this naturally radioactive world, our bodies take in radioactive material through inhalation and ingestion (food and drink) and so we are naturally radioactive (as are other species). Natural radioactive material decays and these decay mechanisms are valuable for dating the rocks and minerals on the Earth (radiometric dating). Professor Robert White, a scientist and Christian, in his paper *The Age of the Earth*, explains how radiometric dating supports the antiquity of the Earth at 4,566 million years old.

19. Table compiled from data in Davis et al., "Oklo Reactors," 3. The authors give the 4.0 percent enrichment for uranium-235 occurring 2.1 billion years ago and point out the small contribution from uranium-234 in today's uranium ore.

Conclusion

In this chapter we have seen that Genesis presents God as the creator of the heavens and the Earth. Christian theology teaches this and Christians believe it. God created the universe including our sun, stars, and the Earth. His creation in the book of Genesis is described as "good" and "very good."

This chapter introduced the current basic scientific understanding of our universe with a focus on its necessary nuclear resources and natural radioactivity. We now know, through modern science, that the universe is powered by nuclear energy—nuclear fusion in the sun and the stars. This nuclear energy provides us with heat and light. Through nuclear fusion in the stars the chemical elements were created that make up our universe, Earth, and our bodies. Nuclear fission and natural nuclear reactors are also part of God's good creation. Moreover, our universe and Earth are naturally radioactive, and so are we human beings, alongside other species. Within this good creation human beings have evolved. Nuclear fusion in the sun gives us daylight while nuclear fusion in the stars lights the night sky. That's worth remembering as we discuss these things.

Of course, the ancient author(s) of Genesis did not know that God's created universe was nuclear powered and radioactive or that the heat and light we receive from our sun is from nuclear fusion. Nor did they know that the light we receive from the moon is a reflection of the sun's (nuclear-generated) light. But through modern science we know this today. The important issue here is that what God created is good and continues to be since he sustains it. When we debate nuclear power we must consider this within the widest possible context, which is God's universe.

Christians, and non-Christians, should take this vital perspective into account. In the next chapter we consider radioactivity and radiation exposure in more detail, examining naturally occurring radioactivity and artificial radioactivity together, with the radiation doses that human beings receive from exposure to these sources. Once again, this will assist us to view nuclear power plants within their correct context.

Further Reading

Alexander, Denis. *Rebuilding the Matrix: Science and Faith in the 21st Century*. Oxford: Lion, 2001.

Done, Chris. *Thinking About . . . the Big Bang: How Did the Universe Begin?* Thinking About . . . leaflet series. Christians in Science, 2016. http://www.cis.org.uk/resources/thinking/.

Holder, Rodney D. *Is the Universe Designed?* Faraday Paper 10. Faraday Institute for Science and Religion, St. Edmund's College, University of Cambridge, April 2007. http://www.faraday.st-edmunds.cam.ac.uk/resources/Faraday%20Papers/Faraday%20Paper%2010%20Holder_EN.pdf.

———. *Thinking About . . . Fine Tuning: What Does It Mean for Our Universe to Be Fine-Tuned?* Thinking About . . . leaflet series. Christians in Science, 2016. http://www.cis.org.uk/resources/thinking/.

Hutchings, David, and Tom McLeish. *Let There Be Science: Why God Loves Science, and Science Needs God*. Oxford: Lion, 2017.

Lucas, Ernest. *Interpreting Genesis in the 21st Century*. Faraday Paper 11. Faraday Institute for Science and Religion, St. Edmund's College, University of Cambridge, 2007. http://www.faraday.st-edmunds.cam.ac.uk/resources/Faraday%20Papers/Faraday%20Paper%2011%20Lucas_EN.pdf.

McLeish, Tom. "Faith and Wisdom in Science." Video interview with Eleanor Puttock at the Faraday Institute Summer Course No 9: Science and Religion – Engaging in Constructive Dialogue, October 30, 2014. https://www.youtube.com/watch?v=C8X6p17TTQI.

Weaver, John. *Christianity and Science*. London: SCM, 2010.

White, Robert S. *The Age of the Earth*. Faraday Paper 8. Faraday Institute for Science and Religion, St. Edmund's College, University of Cambridge, 2007. http://www.faraday.st-edmunds.cam.ac.uk/resources/Faraday%20Papers/Faraday%20Paper%208%20White_EN.pdf.

Questions

1. How has understanding the universe as nuclear powered (through nuclear fusion) and naturally radioactive helped to provide a wider perspective on: (a) understanding God's creation, and (b) issues surrounding the operation of nuclear power plants?

2. How has the discovery of natural uranium fission reactors which operated in Oklo (West Africa) almost two billion years ago helped you understand more about our world and the context for considering modern nuclear reactors?

2

Radioactivity and Radiation Exposure on Earth

> ... Faraday stands firmly in a tradition of natural philosophers like Kepler, Boyle, Ray, Newton, Pascal and many others, who took the biblical content of their faith seriously and who showed a particular delight in uncovering the regularities that characterized God's world.
>
> DENIS ALEXANDER[1]

> Most people's everyday perception of radioactivity is often shaped by the debate on how safe or risky nuclear power plants or nuclear waste repositories are.... It is only when the presence of material in their personal living environment is brought to their attention that they realise that besides radioactive materials from nuclear industries or medical treatment, other forms of radioactive substances also exist.
>
> CLAUDIA KÖNIG ET AL.[2]

---

1. Alexander, *Rebuilding the Matrix*, 164. Dr. Denis Alexander is a scientist and Christian. He is Emeritus Director of the Faraday Institute for Science and Religion at St. Edmund's College, University of Cambridge.

2. König et al., "Remediation of TENORM Residues," 576. The authors worked at the Institute of Radioecology and Radiation Protection, Leibniz Universität, Hannover, Germany.

Introduction

IN CHAPTER 1 WE saw that naturally occurring nuclear fusion, radioactivity, and nuclear fission form the fabric of God's created universe. Now we need to look more at Earth. Many people's perception of radioactivity and radiation exposure is just focused on the context of civil nuclear power plants, disposal of radioactive waste, and possibly from medical procedures. In fact, it is more complicated than that, as we will see. People are exposed to both naturally occurring radioactivity and artificial radioactivity. For example, the air we breathe and the food we eat are naturally radioactive.

Besides these natural sources of radioactivity, we are routinely exposed to artificial (anthropogenic) sources of radiation. For example, most people will think about medical/dental procedures exposing them to x-rays, but, as we will see, there are other sources too.

Therefore, this chapter outlines the sources of exposure, and gives the average annual doses that the UK population receives. These doses are compared with population doses in our neighbor Ireland and the US. Understanding these radiation sources and the doses received from them provides a constructive context for understanding actual and potential exposures of people from nuclear plants and its implications. This approach puts the nuclear energy discussion in a valuable wider social perspective on risks (looked at further in chapter 3). Recently, I reviewed two books on New Testament studies and they both thoroughly addressed the historical, social, and cultural contexts in which the New Testament was written. Context is critical for the helpful interpretation of facts.

Countries conduct periodic reviews of their population doses to identify any changes over time. In the UK periodic reviews have been made since 1974 by the National Radiological Protection Board (NRPB), then the Health Protection Agency (HPA), and now by Public Health England (PHE). These examine all significant exposures. The review by Watson and colleagues in 2005 considered data from 2001 to 2003.[3] They noted that there was little change over the years.

The most current review (the eighth, from 2010) by PHE compares its results with the 2005 review. For each exposure source it provides the estimated total population dose (called the collective dose) and the per caput dose (i.e., the average individual dose, which is the term I will use).[4] The

---

3. Watson et al., *Ionising Radiation Exposure* (2005).
4. Their term dose is shorthand for "effective dose" as defined by the International

2005 review addressed population exposure under the two main categories: natural and artificial. However, the 2010 review uses two slightly different categories: (1) ubiquitous radiation in the environment, and (2) exposure from the use of radiation.[5] We follow these below.

Exposure of the UK Population from Radiation in the Environment

Some people are worried about radioactivity and radiation exposure from nuclear energy, but have no context to understand such things. We have already seen that our universe and world are naturally radioactive. The sun and space expose the Earth, and its inhabitants, to cosmic radiation; the air we breathe is naturally radioactive, and so are our food and drink. The Earth's oceans, soils, and rocks are radioactive and so are the building materials used in our homes. Even our bodies are naturally radioactive! We cannot escape exposure to natural radioactivity. The 2010 review category "ubiquitous radiation" includes natural radioactivity plus widely distributed sources in the environment from weapons fallout and other anthropogenic sources. The six categories that expose populations are: (1) radon and thoron gas, (2) intakes of natural radioactivity (apart from radon/thoron), (3) terrestrial gamma radiation, (4) cosmic radiation, (5) weapons fallout, and (6) other anthropogenic sources.

Imagine an Invisible Gas

Imagine a gas that is naturally occurring, colorless, tasteless, and odorless. It is everywhere on the Earth and we cannot identify it with our senses. What is it? Of course, you may think of the atmosphere and say oxygen or nitrogen (and you would be right), but I am thinking of a gas that, besides having the above properties, is also radioactive. And this is not something

---

Commission on Radiological Protection (ICRP). Effective dose takes account of the different types of radiation exposing a person and also the organs and tissues irradiated (with tissue weighting factors being used for organs and tissues). Balter et al., "Radiation Is Not the Only Risk," 763, has a good explanation, or see my glossary.

5. Oatway et al., *Ionising Radiation Exposure*, iii. The Executive Summary table uses these terms, as does table 25 (p. 34), etc. Note that this 2010 reivew was published in April 2016. It is the latest UK review. The 2005 review (Watson et al.) used the more common headings of "natural radioactivity" and "artificial radioactivity." The term "ubiquitous radiation" refers to radiation that is widespread in our environment.

from science fiction, another planet, or a remote galaxy. It is science fact. This gas exists on the Earth and it is dangerous. It is not confined to scientists' research laboratories, secret government research projects, or even nuclear plants. This gas is found in our workplaces and—even worse—in our homes. It is scary—naturally occurring, colorless, tasteless, odorless, radioactive, and everywhere. But there is more: this gas is a known killer. As we cannot detect it with our senses we need scientific equipment to detect its presence.

This ghastly gas is radon (Ra-222). It comes from tiny levels of natural uranium in the soils, rocks, and building materials. As radioactive uranium undergoes decay it produces further radioactive products, including the radioactive gas radon. This becomes airborne and inhaled. Radon's radioactive decay products are also inhaled and these too irradiate our lungs. The exposure to radon gives a radiation dose to our lungs increasing the risk of us getting lung cancer (see chapter 3). Radon gas is internationally recognized as a public health hazard. Actually, radon gas is not alone, although it is our main exposure source. Another gas, radon-220 (Ra-220), generally called thoron, is released during the decay of naturally occurring thorium. It also gives us a radiation exposure of our lungs.

PHE provides the UK's primary resource for essential information on radon covering the public, schools, workplaces, and professionals. It remarks that outdoor radon concentration is low, but inside buildings its concentration increases. So, radon enters our homes, schools, and workplaces, where it may concentrate in hazardous levels depending upon a number of factors. For example, in UK dwellings radon concentration: (1) is higher in the winter months and lower in the summer, (2) varies with the days of the week, (3) varies with the time of day (dependent upon opening/closing doors and windows), and (4) depends on the region (geology) where you live.[6] Even when we are outdoors radioactive radon is still present. People, thankfully, are becoming more aware of its presence through public education programs and official measures, e.g., radon surveys of homes.

About 48 percent of the UK's average individual annual dose comes from radon and thoron gas. This is a dose of 1.3 mSv per year (pronounced 1.3 millisieverts per year).[7] But it is just an average; some people are ex-

6. See PHE, "Radon at a Glance," under the heading "What Is Radon?"

7. Ibid. The average is calculated assuming that the airborne concentration of radon is 20 Bq m^{-3} (or 20 Bq/m^3) in England (pronounced 20 becquerels per cubic meter). 1 Bq is one nuclear disintegration per second. Further, 1 mSv = 1/1,000 Sv. A sievert is a large unit, so millisieverts are used. These are the international (SI) units.

posed to higher levels. For example, in Cornwall the average annual dose is almost 7 mSv.[8] The UK's average airborne indoor concentration of radon is 20 Bq m^{-3}. This gives an effective dose of 1 mSv per year. Outdoor airborne concentration of radon and thoron is lower at 4 Bq m^{-3}.

Because around 1,100 deaths occur on average annually in the UK from radon-induced lung cancer, we must realize that deaths also occur in other countries through radon exposure. For example, in the US, *The National Radon Action Plan* states: "Our ultimate goal is to eliminate avoidable radon-induced lung cancer in the United States..."[9] Radioactive radon is ubiquitous and a major public health risk worldwide. Nevertheless, many people remain unaware of its reality and risk in their own homes.

Intake of Natural Radionuclides: Food and Drink

In addition to breathing in radon and thoron gas, we also take natural radioactive material into our bodies through ingestion. Food is naturally radioactive and so is water![10]

For example, carbon-14 (C-14, half-life 5,730 years) is produced in the atmosphere from cosmic radiation interacting with nitrogen. It then enters plants and so comes into the human food chain as we eat the plants. However, other naturally occurring radionuclides are also ingested in our food. Radioactive potassium-40 (K-40, half-life 1.3 billion years) occurs as a small percentage by weight in all potassium on Earth. It is taken into our bodies, which maintain a fairly constant level of potassium-40.

Radionuclides in the natural uranium and thorium series also irradiate the population, with lead-210 (Pb-210) and polonium-210 (Po-210) making the main contributions. For example, the population is exposed from consumption of vegetables, meat, dairy products, fish, and cereals. Drinking water is another route for exposure via natural radioactivity in water, including radon.[11] The average dose estimated from intake of natu-

---

8. Oatway et al., *Ionising Radiation Exposure* (2010), 3, fig. 1.
9. American Lung Association et al., *National Radon Action Plan*, 2.
10. Cefas, *Radioactivity in Food and the Environment*, 2014.
11. In 2013 the European Commission published a Council Directive on limits of radioactive substances in drinking water. It became UK law in 2015. In 2014 the UK's Food Standards Agency (FSA) surveyed bottled water and concluded: "The average annual background dose received by people living in the UK is not significantly increased by high-rate consumption of any of the bottled waters." FSA, *Radioactivity in Bottled Water*, 1.

rally occurring radionuclides (excluding radon and thoron gas) is 0.27 mSv per year (11 percent of the total average dose).

Some foods can contain higher levels of natural radioactivity. Brazil nuts, for example can have enhanced levels of radium isotopes (e.g., radium-226, half-life 1,600 years) and the 2010 PHE review notes that eating 100 g of Brazil nuts gives a dose estimate of 0.01 mSv. Finally, the 2010 review noted that just smoking one cigarette per day for a year gives a dose of about 0.018 mSv (mainly from lead-210 and polonium-210 in tobacco leaves, which are not removed during manufacture).

Terrestrial Gamma Radiation

Another source of natural radiation exposure is from gamma radiation coming from rocks and soils. This is a terrestrial source. The tiny levels of radioactive uranium (and thorium) that are naturally present in the Earth's soil and rocks undergo radioactive decay over long periods of time into what are called decay products; in the process they give off emissions of gamma rays that irradiate us.[12] Radioactive potassium-40 is also present.

The exposure we receive depends on the concentration of the radioactive materials in rock formations and the building materials used for our homes. So these gamma radiation exposures contribute to what is called the natural background dose. UK surveys have established the level of exposure from terrestrial gamma radiation. The average annual dose estimate (from time spent indoors and outdoors) is about 0.35 mSv (13 percent of the total average dose).

Cosmic Radiation

Another source of our background dose is from naturally occurring cosmic radiation that bombards the Earth from the sun and space. These very energetic particles (mainly protons, alpha particles, and electrons) hit our atmosphere, collide with atoms, and give secondary radiation (e.g., neutrons are detected at ground level).[13]

---

12. These elements are called primordial radioisotopes, which existed from when the Earth was formed.

13. Alpha particles are made of two protons and two neutrons, so they are heavy and positively charged. They are the same as a helium-4 nucleus.

Thankfully, Earth's atmosphere provides significant shielding for those living at sea level. However, if you live at higher altitudes then you will be exposed to more cosmic radiation than somebody living at sea level. In the UK this difference is small. Our buildings provide some shielding from cosmic radiation. Taking account of time spent indoors and outdoors, the UK average dose estimate is 0.33 mSv per year (12 percent of the total average dose).

Nevertheless, the difference in dose rate is significant between sea level and altitudes for commercial aircraft flights. Aircrew and passengers are subjected to higher exposure levels from cosmic radiation during their flights. Bartlett and colleagues remarked: "Aircraft crew and frequent flyers are exposed to elevated levels of cosmic radiation of galactic and solar origin and secondary radiation produced in the atmosphere, the aircraft structure and its contents."[14] Again, based on assumptions (e.g., number of passengers, their destinations, and an average dose rate in the airplane of 0.004 mSv per hour), the average annual dose is estimated as 0.03 mSv per year. Clearly, those who don't fly as passengers have a zero dose from this source while those who do fly are exposed. The average is determined from the whole population. Interestingly, if you do not fly or you reduce your flying hours to reduce your carbon footprint, you receive the additional benefit of also reducing your radiation exposure![15] Aircrew doses are considered under employment exposures below.

Nuclear Weapons Fallout

The 2010 review includes radiation exposure from atmospheric nuclear weapons tests conducted by various countries (including Britain) which occurred in the 1950s and 1960s. It was a component of the Cold War. A test ban treaty signed in 1963 considerably reduced atmospheric testing, so most fallout occurs from detonations made in the early 60s. These tests released

---

14. Bartlett et al., *Health Protection Agency* (2006), i. Further, in 2016 Alvarez et al., "Radiation Dose to the Global Flying Population," provided a detailed analysis of doses to passengers. The authors are from the Laboratory for Aviation and the Environment, Department of Aeronautics and Astronautics, MIT, US. They demonstrated that it was possible, in some circumstances, for frequent flyers to exceed the International Commission on Radiological Protection annual dose limit for members of the public (1 mSv/year).

15. For example, Climate Stewards, *London–New York Return*, calculates that this trip of 3,442 miles results in 2.5 tons of CO_2 being emitted per person, and the carbon offset is £38. Climate Stewards is part of the A Rocha network, a Christian organization.

radioactive materials high into the atmosphere, which then entered our air and rainwater, and were deposited on the Earth. Although decades of radioactive decay have reduced the doses received, some important radionuclides are still in our environment. Cesium-137 irradiates populations through its external gamma radiation. Carbon-14, strontium-90, and tritium (hydrogen-3) are taken into food and then ingested. The average annual dose is estimated at about 0.005 mSv (0.2 percent of the total average dose).

Other Anthropogenic Radioactivity

The 2010 review identifies two other sources of ubiquitous radioactivity: discharges and accidental releases. The UK's civil nuclear sites discharge to the atmosphere and marine environments from four processes: fuel manufacture, reactors, fuel reprocessing, and radiopharmaceutical production. The average annual individual dose is 0.0002 mSv. Non-nuclear industries also discharge to the environment, but incomplete records mean an estimate was not derived, although the review reckoned the population doses received were lower than discharges from the nuclear industry.

Accidental releases from the UK's 1957 Windscale fire and the Chernobyl accident in Ukraine in 1986 are discussed, but the UK population's collective dose "is expected to be insignificant." The Fukushima Daiichi accident in Japan that occurred in March 2011 is outside the period of the 2010 review and will be considered in the next review.

Consequently, the 2010 review estimated an average dose of just 0.0008 mSv to other anthropogenic radioactivity (0.01 percent of the total average dose).[16]

Exposure of the UK Population from Using Radiation

The second category in the 2010 review is "Exposure from the use of radiation," subdivided into (1) patient exposure from medical use and (2) occupational exposure from the use of radiation (meaning people exposed in employment).

---

16. Oatway et al., *Ionising Radiation Exposure* (2010), 10–11.

Patient Exposure

Medical exposure of patients contributes 16 percent of the average annual dose. What are its sources? These are diagnostic and therapeutic procedures. Diagnostic radiology includes CT (computerized tomography), conventional radiology, angiography (non-CT), and interventional (non-CT). CT examinations give higher doses than other procedures and over the years its contribution has risen to about 68 percent of the collective dose from diagnostic radiology. The "total diagnostic radiology" average annual dose was 0.40 mSv. In 2008 (the last year for data) the number of UK medical and dental examinations was estimated at about 46 million.

Nuclear medicine procedures (where radioisotopes are used inside the body) gave an average annual dose of 0.03 mSv.

In therapeutic procedures involving radiotherapy and nuclear medicine, the target organ(s) are treated but non-target organs may also receive some radiation exposure. This exposure leads to an estimated 0.01 mSv to the UK's average individual dose.

Added together, the average annual dose to the UK population from all medical uses is estimated at 0.44 mSv.

Occupational Exposure

Besides medical exposure of patients there are people who receive radiation exposure during their employment. This is called occupational exposure. The 2010 review includes people employed in the nuclear industry, the gas and oil industry, medical workers, veterinary workers, the defense industry, radionuclide production industry, general industry, research and tertiary education, and transport of radioactive materials. The average annual dose from occupational exposure was estimated at 0.0004 mSv—a very low dose.[17]

Aircraft crews come into the occupational category from their enhanced exposure to cosmic radiation during passenger flights. The average annual individual doses to aircrew was estimated as 2.4 mSv (similar to the previous review).[18] Readers may be surprised that in 2010 this figure of 2.4 mSv exceeded the average doses received by people exposed in the nuclear industry.

---

17. Ibid., 34. See their Table 25, which shows exposure to the UK population from all significant radiation sources.

18. Ibid., 7.

Comparing the exposures in 2003 and 2010, the 2010 review stated:

> The most significant change in any exposure ... was that, between 2003 and 2010, the per caput dose to the UK population from patient exposure during CT examinations increased from about 0.18 mSv to about 0.27 mSv. In 2010, patient exposure during CT examinations accounted for about 10% of the per caput dose to the UK population, or about 62% of the dose from all medical examinations.[19]

Exposure from Using Consumer Products

Other sources of exposure for the population include consumer products, e.g.: (a) radioactive americium-241, a small source used in ionization chamber smoke detectors, located in homes, business premises, and shops storing smoke detectors for sale; and (b) wrist watches containing radioluminous material such as radioactive tritium (H-3). In the 2005 review the combination of consumer products amounted to an average UK population dose of 0.0001 mSv per year (less than 0.1 percent of doses). However, the 2010 review commented that there was insufficient data on the number of people exposed to consumer products to determine their contribution to the average annual dose.[20]

Table 2 summarizes the UK population doses.

---

19. Ibid., 34.
20. Ibid., 1. See 27–28 for a short discussion.

Radioactive sources	Average dose (mSv)
Ubiquitous radiation in the environment	2.3
(Natural background, weapons fallout, and discharges)	
From: radon and thoron gas (1.3 mSv), intakes of natural radioactivity in food and drink (0.27 mSv), terrestrial gamma radiation (0.35 mSv), cosmic radiation (0.33 mSv), weapons fallout (0.005 mSv), and other anthropogenic radioactivity (0.0008 mSv).	
Exposure from using radiation	0.44
From: patient exposure during medical procedures (0.44 mSv), and people exposed in employment (termed occupational exposure) (0.0004 mSv).	
Total dose	2.7

Table 2: Average Annual UK Population Exposures in 2010[21]

Exposure of Ireland's Population

After discussion of each source for the UK's population, space does not permit me to repeat the procedure for other countries. However, I draw a brief comparison with the UK's neighbor Ireland and then the US. In 2010 the Radiological Protection Institute of Ireland (RPII) and Health Service Executive (HSE) produced a detailed joint position statement on radon gas. This showed that for Ireland the average indoor level was 89 Bq/m³. The World Health Organization (in 2009) compared this level with the world and identified it as "the eighth highest average concentration."[22] This is much higher than the UK's average of 20 Bq/m³, meaning that Ireland's population receives a higher exposure than the UK's. Ireland's population has a significant exposure from radon gas (over 55 percent of the average dose received).

---

21. Ibid. My table is a simplification compiled from the data and tables provided by the authors, especially on their pages iii and 34. For their helpful figure on the UK population dose see page iv.

22. RPII and HSE, *Radon Gas in Ireland*, 1.

In 2014 the RPII published an impressively detailed analysis of Ireland's population doses.[23] A comparison between the 2008 and 2014 assessments shows that the average individual annual dose was 3.95 mSv in 2008 and 4.037 mSv in 2014—the results were consistent.[24] Both figures are higher than the UK's average dose to its population and Ireland does not have any nuclear plants. However, discharges from the Sellafield reprocessing plant in England into the Irish Sea were assessed for Ireland's population, and this showed a steadily reducing contribution of exposures from this source, to a low level in recent years up to 2011.[25]

Exposure of the US Population

As a useful comparison, we can see what the natural and artificial sources are for the US population. Richard Wakeford summarizes the report of the National Council on Radiation Protection and Measurements (NCRP) on the US population doses in 2006 (which compares values to the 1987 report).[26] NCRP's executive director David A. Schauer also gave a clear illustrated overview.[27]

The average ubiquitous natural background radiation effective dose was 3.1 mSv (two thirds coming from radon gas).[28] The main artificial source was medical procedures, giving 3.0 mSv (a considerable rise from 0.53 mSv in the 1987 report). Half of the dose comes from CT scans and one quarter from nuclear medicine. This dramatic rise is a notable finding.

Consumer products and activities gave an average dose of 0.13 mSv (with cigarette smoking alone accounting for 35 percent of this figure). The areas of industrial activities, security, medical procedures, education and

---

23. O'Connor et al., *Doses Received by the Irish Population*, 50, mentions that Ireland's high radon doses were being investigated by the authorities.

24. In Ireland, 86 percent of all radiation exposures arose from natural sources (3.48 mSv), with 14 percent from artificial sources (0.557 mSv, primarily from medical procedures). Ibid., 3, 47. The report has doses in microsieverts (μSv), which I have converted to mSv (1,000 μSv = 1 mSv).

25. Ibid., 29.

26. Wakeford, Review of *NCRP Report No. 160*. Richard Wakeford is a visiting professor at the Dalton Nuclear Institute, University of Manchester (http://www.dalton.manchester.ac.uk) and Editor-in-Chief of the *Journal of Radiological Protection*.

27. Schauer, *Ionizing Radiation Exposure*.

28. Wakeford, Review of *NCRP Report No. 160*. Radon gas gave some people a high effective dose of 11.1 mSv.

research activities gave a population annual average of 0.003 mSv. This included caring for and contact with nuclear medicine patients (72 percent of the dose) and nuclear power generation (15 percent of the dose).

Employment exposure was calculated over the US population to give an average of 0.005 mSv. The percentages received by employment were subdivided into: medical (39 percent), aviation (38 percent), commercial nuclear power plants (8 percent), and industry and commerce (also 8 percent). Two further groups received lower percentages. The highest average occupational doses were for aircrew (3.1 mSv per year).

This analysis gave an average annual US total from natural and artificial sources of 6.2 mSv; the dose from artificial sources is now the same as that for natural background sources. The average of 6.2 mSv is more than double the 2.7 mSv average annual dose received in the UK. In each country there is a range of doses; however, knowing what the average is helps with perspective since we can see if this changes over time by assessing the various components. It also allows us to compare the average background with doses received in other countries. However, the American Institute of Physics (AIP) was concerned that the increase in dose from medical procedures could be misinterpreted, and they issued a cautionary note: "The report is not without scientific controversy and requires careful interpretation." They emphasized the benefits of medical procedures.[29]

Radioactive sources	*Average dose (mSv)*
Ubiquitous natural background	3.1
From: radon gas, gamma radiation, cosmic radiation, and food/drink.	
Artificial radiation	3.13
From: medical exposures; consumer products (including smoking); industrial, security, and other activities; and occupational exposure.	
Total dose	6.23 (rounded to 6.2)

Table 3: Average Annual US Population Exposures in 2006[30]

29. AIP, "Radiation Exposure of U.S. Population." Also, see Balter et al., "Radiation Is Not the Only Risk."

30. Table compiled from data in Wakeford, Review of *NCRP Report No. 160*.

Further Information on Artificial (Anthropogenic) Sources

Hospital Discharges

Radioactive discharges are rightly of concern to many people, including legislators and the general public. However, besides nuclear installations, it is important to recognize that non-nuclear establishments such as hospitals are routinely involved in radioactive discharges as permitted by legislation.[31] For example, McGowan and colleagues commented:

> Hospitals and other operators discharge radioactivity into the sewers as a result of excretions from patients undergoing diagnostic and therapeutic nuclear medicine procedures. Such discharges usually flow directly to sewage treatment plants, and there is an ongoing debate on the benefit of local abatement systems to reduce such discharges.[32]

They studied three UK hospitals where liquid discharges went via the sewage plants and then other watercourses, eventually into the River Thames. Doses to workers and members of the public were assessed. These showed that the doses received were low and did not merit additional costs to the public. As this careful study demonstrated, iodine-131, which is used to treat cancer patients, results in discharges to the environment yet the low doses received by workers and members of the public did not justify additional expenditure on abatement tanks which allow discharged iodine-131 to decay.

Naturally Occurring Radioactive Material (NORM)

Naturally occurring radioactive material exposes us daily, as discussed above. However, more discussion here will help us understand the issues involved as we look at examples. Some industrial processes result in the concentration of NORM, which can expose workers and members of the public to additional levels of radiation. This is called industrial NORM, or TENORM (technologically enhanced NORM). Read and colleagues warned: "NORM is an unfortunate and misleadingly reassuring acronym for such contamination; only the nuclides are 'naturally occurring'; the materials themselves

---

31. For hospital discharges, HMG's *Environmental Permitting (England and Wales) Regulations 2010* (as amended) apply.

32. McGowan et al., "Iodine-131 Monitoring," 1. D. R. McGowan worked in Radiation Physics and Protection, Oxford University Hospitals NHS Trust, Oxford.

are the products of chemical processing and can be extremely hazardous."[33] Nevertheless, that's the term used so we'll need to use it.

Read and colleagues helpfully discussed industrial NORM and identified current standards and regulations. For example, to address NORM the International Atomic Energy Agency (IAEA) updated its International Basic Safety Standards in 2011, with the European Commission revising its Euratom Basic Safety Standards in 2012.[34] The IAEA Basic Safety Standards identified "NORM industries" (e.g., those producing certain fertilizers) and in the UK the legislators have added china clay (see below). These NORM industries may result in the production of radioactive waste, contaminated land, exposure of workers, and exposure of the general population.

First, there is the issue of nuclear waste—a frequently raised objection to building new nuclear power stations. But to put this in some perspective, Read and colleagues observed:

> The volume of radioactive waste created each year by the nuclear industry is dwarfed by that from other sources, notably the production and combustion of fossil fuels and the exploitation of industrial minerals (IAEA 2003). Some of the scales affecting production equipment contain activity levels corresponding to those of intermediate level nuclear waste (ILW) . . .[35]

However, they rightly remark that most of the waste is much less active than ILW, but it still must be dealt with and disposed according to the legislation.

Read and colleagues also addressed the ongoing production of china clay (kaolinite) in the UK counties of Devon and Cornwall, an industrial process originating over two hundred years ago. On one Cornwall site, scales deposited inside equipment were radioactive from NORM and some soil was similarly contaminated, meaning that remedial actions were taken to clear the NORM contamination.

A different example of problems caused by industrial NORM is from residential dwellings built on contaminated land in two cities in Germany. High contamination from toxic chemicals and radioactive material from previous industry meant that remediation was needed to reduce risks to

---

33. Read et al., "Background in the Context," 369. D. Read worked in the Department of Chemistry, Loughborough University, UK.

34. Ibid., 368.

35. Ibid., 367. They referenced IAEA, *Extent of Environmental Contamination* (2003). Scales are deposits on the industrial equipment.

the general public: contaminated soil had to be removed from the sites as waste. Concerns over risks and communication were ongoing challenges.[36]

NORM is also an issue in extracting gas and oil from shale via hydraulic fracturing ("fracking"). For example, Cefas commented on the exploration and extraction of shale gas in the UK: "This process, along with others for unconventional sources of gas such as coal bed methane, represents a potential source of exposure of the public and workers."[37]

PHE has issued a review of this potential hazard. Moreover, an article by Joel Garner and colleagues has demonstrated the presence of NORM in the current oil and gas industry in two onshore production sites.[38] They noted that this is common in offshore oil and gas production in the North Sea, but their paper identified it as an onshore issue also, in addition to one production site previously identified in Dorset. Their work is relevant if fracking is started in the East Midlands.

Radioactivity in the UK's Food and Environment

In this chapter we have considered the UK's population exposure from artificial sources of radioactivity, including radioactive waste and discharges. Much more information on discharges is readily available online as reported by the Environment Agency (EA) and the Food Standards Agency (FSA).[39] This report is a compilation of results for multiple government environmental agencies and food standards agencies, and Natural Resources Wales. It addresses England, Wales, Scotland, and Northern Ireland, covering: nuclear fuel production and reprocessing, research establishments, nuclear power stations, defense establishments, radiochemical production, industrial and landfill sites, regional monitoring, etc. It is a massive report providing detailed data for individual sites.

As we are looking at nuclear power plants, I will just say a little on the information available here. Take, for example, Berkeley and Oldbury Power Stations (north of Bristol, situated on the River Severn).[40] These are two Mag-

---

36. König et al., "Remediation of TENORM Residues." TENORM (sometimes written as TNORM) is technologically enhanced naturally occurring radioactive material.

37. Cefas, *Radioactivity in Food and the Environment, 2014*, 29–30.

38. Garner et al., "NORM in the East Midlands." Joel Garner, Department of Chemistry, Loughborough University, Loughborough, UK.

39. Cefas, *Radioactivity in Food and the Environment, 2014*.

40. I selected Oldbury and Berkeley as I worked at Oldbury Power Station and also

nox stations now being decommissioned. The data provided gives key points, and total doses to the public from all exposure pathways (less than 0.005 mSv in 2014, slightly down on the 2013 value 0.10 mSv).[41] However, a new assessment to houseboat dwellers from external exposure cautiously estimated 0.022 mSv and so this was not included in the total dose at this stage.

The report discusses gaseous discharges with terrestrial monitoring and liquid discharges with aquatic monitoring giving: (1) concentrations of radioactivity in food and the environment (both marine and terrestrial samples), and (2) radiation dose rates near the power stations. So here is an excellent freely-available authoritative reference, which is readable and, I think, reassuring in the measurements taken and the analysis provided. The statutory UK annual dose limit to members of the public is 1 mSv per year, so 0.005 mSv is only 0.5 percent of this limit. Pretty small! Even with the preliminary estimate of 0.022 mSv to houseboat dwellers we have low annual doses.

This report also mentions overseas incidents and controls of UK food imports following the 1986 Chernobyl accident and the Fukushima nuclear plant accident in Japan in March 2011.[42]

In October 2016, an updated report provided results for 2015. For the two decommissioning Magnox power stations at Berkeley and Oldbury, mentioned above, it reported total doses to the public from all exposure pathways (less than 0.005 mSv in 2015; unchanged from the 2014 value). Now, as in 2014, 0.005 mSv is just 0.5 percent of the dose limit (1 mSv per year). The estimated dose to houseboat dwellers from external exposure was 0.006 mSv (down from the 2014 estimate of 0.022 mSv).[43]

The above reports consider radioactivity in food. However, it is worth mentioning that sometimes food is irradiated before sale to extend its shelf life and kill bacteria (e.g., salmonella) that cause food poisoning. The food doesn't become radioactive and it is safe to eat. Legal controls are in operation covering the practice, labelling, distribution, etc.[44]

Oldbury Training Centre for many years. Oldbury and Berkeley are close together and included as one site in the report.

41. Cefas, *Radioactivity in Food and the Environment*, 2014, 110–11.
42. Ibid., 14.
43. Cefas, *Radioactivity in Food and the Environment*, 2015, 110–11.
44. For further information, see FSA, "Irradiated Food."

Conclusion

In the UK, sources of ionizing radiation to the general public are periodically reviewed. A review in 2005 showed that the average annual radiation background dose to the UK's population from all significant radioactive sources, natural and artificial, was 2.7 mSv.[45] The 2010 review also estimated the exposure at 2.7 mSv. We know that there are ranges so exposure does vary. Contributions to the average annual UK population dose from the nuclear industry are very small.

Ireland's population has a significant exposure from radon gas (over 55 percent of the dose received). The average individual annual dose is 4.037 mSv even though Ireland does not have any civil nuclear plants.

For the US population the average radiation dose from natural sources is 3.1 mSv, while the main artificial source is medical procedures, giving another 3 mSv. This gives a total average annual radiation dose of 6.2 mSv. Most natural exposure occurs from radon gas and most artificial exposure comes from medical practices.

There are risks and benefits in being exposed to radiation and these are introduced in the next chapter, along with discussion on reducing risks. Finally, although we have focused on human exposure to radiation, we note that other living creatures on Earth are exposed to radiation, but considering this is beyond the scope of this chapter.

Further Reading

American Lung Association, et al. *The National Radon Action Plan: A Strategy for Saving Lives*. EPA (US), November 2015. https://www.epa.gov/sites/production/files/2015-11/documents/nrap_guide_2015_final.pdf.

Environmental Protection Agency (US). *Basic Radon Facts*. EPA 402/F-12/005. February 2013. https://www.epa.gov/sites/production/files/2016-08/documents/july_2016_radon_factsheet.pdf.

Oatway, W. B., et al. *Ionising Radiation Exposure of the UK Population: 2010 Review*. PHE-CRCE-026. PHE, April 2016. https://www.gov.uk/government/publications/ionising-radiation-exposure-of-the-uk-population-2010-review.

O'Connor, C., et al. *Radiation Doses Received by the Irish Population 2014*. RPII 14/02. RPII, June 2014. http://www.epa.ie/pubs/reports/radiation/RPII_Radiation_Doses_Irish_Population_2014.pdf.

---

45. Cefas, *Radioactivity in Food and the Environment*, 2014, 30. They cited Watson, *Ionising Radiation Exposure*. Watson's section 8 discussion (pp. 70–80) is valuable. See also Martin et al., *Introduction to Radiation Protection*, 42–45.

Radiological Protection Institute of Ireland, and Health Service Executive. *Radon Gas in Ireland: Joint Position Statement by the Radiological Protection Institute of Ireland and the Health Service Executive*. April 2010. http://www.epa.ie/pubs/reports/radiation/RPII_Radon_Ireland_Joint_HSE_10.pdf.

United Nations Environment Programme. *Radiation: Effects and Sources*. 2016. http://www.fs-ev.org/fileadmin/user_upload/89_News/Oeff.-Arbeit/Radiation_Effects_and_sources-2016.pdf.

Questions

1. Can you calculate your estimated personal annual radiation dose? Use the link:

    American Nuclear Society, "Radiation Dose Calculator." February 8, 2016. http://www.ans.org/pi/resources/dosechart/.

    There are three choices: (a) selecting from the options shown gives an interactive calculation in earlier units (millirems), (b) for international units (millisieverts) click on that link, and (c) a printable version is available (in millirems) via a link. If you moved to another part of the country how could this change the dose received?

2. For other countries adapt the above website or find a similar one for that country.

3. Watch the video: Public Health England. "Radon, What Can I Do?" (https://www.youtube.com/watch?v=zI6FRrA23cE). How does this help you?

3

Recognizing Risks and Benefits

...the risks that kill you are not necessarily the risks that anger and frighten you...

PETER M. SANDMAN[1]

Risk has a cognitive aspect (i.e. what we know about the risk) and an emotional aspect (i.e. what we feel in terms of dread or fear about it). Until relatively recently health and environmental threat communications have tended to focus on the cognitive aspects (on the assumption that people are rational actors once provided with relevant information), whereas research consistently shows that individuals' actions can be driven by the emotional aspects of risk.

DAVID HEVEY[2]

Introduction

CHAPTER 2 INTRODUCED THE sources of natural and artificial radiation exposure and the average annual population doses received from them. Understanding these sources and doses provides a constructive context for

---

1. Sandman, "Risk Communication," 21. Peter Sandman is a risk expert. See his website at http://www.psandman.com.

2. Hevey, *Review of Public Information Programmes*, 14. Professor David Hevey is Director of the Research Centre for Psychological Health, School of Psychology, Trinity College, Dublin.

considering people's exposures to harm from nuclear plants. This approach puts the nuclear debate and dialogue in a wider perspective.

Nuclear energy must be seen within the context of the benefits and risks of other forms of electricity generation, alongside other risks in life. Indeed, *The Guardian* writer John Vidal highlights a 2016 World Health Organization (WHO) report that concluded 12.6 million people were killed globally (in 2012) through environmental risks. The two major environmental killers were strokes and heart disease.[3] However, people often think of cancers when they think of nuclear energy, so what does the WHO report tell us here? Well, lung cancer is the largest contributor to deaths from cancer, and it is caused by: household air pollution, ambient air pollution, residential radioactive radon, occupational risks, and second-hand tobacco smoke.[4] Air pollution constitutes a major risk to public health, as a 2016 World Bank report confirms.[5]

Risks in Everyday Living

Everyday living brings risks. When I was a teenager in the 1960s I walked, camped, and cycled in the countryside and city. Road traffic accidents were real hazards, but cycle helmets weren't available. At the seaside we risked drowning and sunburn, unaware of the dangers of skin cancer.[6]

At home we had risks from burning coal gas for cooking and using open coal fires for heating. Liverpool was a smoky industrial city with smog. As children we were exposed to fumes, soot, sulfur dioxide, etc., from all the coal burning. The terrible London smog of December 1952 killed 12,000 people from lung conditions. However, the Clean Air Act (1956) began to tackle this terrible pollution from burning coal. Adults, too, were exposed to these hazards. Adults also consumed alcohol and smoked cigarettes for pleasure.

---

3. Vidal, "Environmental Risks Killing" (March 15, 2016). WHO defines environmental risks to health as: "all the physical, chemical and biological factors external to a person, and all related behaviours, but excluding those natural environments that cannot reasonably be modified." Prüss-Ustün et al., *Preventing Disease*, x.

4. Prüss-Ustün et al., *Preventing Disease*, 46–47. Lung cancer resulted in almost 1.6 million deaths (in 2012) and an estimated 36 percent is from environmental factors (ibid., 50).

5. World Bank, and Institute for Health Metrics and Evaluation, *Cost of Air Pollution*. Also see IEA, *Energy and Air Pollution*, and "Presentation."

6. See Prüss-Ustün et al., *Preventing Disease*, 78–79. In 2012 around 372,000 people drowned globally, with children ages 1–4 suffering the most.

Cigarette smoking was the norm in private homes, public places, and on public transport. There was little escape from this pollution apart from in classrooms and churches. Yes, babies, children, and adults were subjected to second-hand smoke almost everywhere daily. Today there are improvements, but cities still suffer from airborne pollution and associated health risks. Now we know that smoking is a serious health threat, causing cancers and shortening lives, so legislation and education help to protect people.

Then there were accidents. I recall the major accident at Aberfan, South Wales, on October 21, 1966. A coal tip—waste created from local coal mines—unexpectedly moved down the mountain and crushed the village's Pantglas Junior School, where lessons were beginning. Mud and debris engulfed classrooms, a farm, and houses. Tragically, 116 children and 28 adults died—only a few children escaped. It was a heartbreaking national tragedy.[7] The tribunal found the National Coal Board responsible for multiple failures in storing the industrial waste (there wasn't a policy on tipping the waste), and it was legally required to pay compensation. Lessons were identified, recommendations made, and future legislation mentioned.[8] This helped to make future waste storage safer, but it could neither bring back the dead nor heal suffering survivors. Besides this, UK coal miners suffered deaths and injuries from accidents and employment-related diseases.[9] Let us now return to alcohol and tobacco.

Alcohol and Smoking: Benefits and Risks

Why start here? Well, people are aware that radiation can induce cancers, but what is not so well known is that alcohol consumption carries cancer risks. Drinking alcohol, and smoking, are considered to be enjoyable and sociable. These are perceived benefits, but the government issues guidelines for sensible drinking to reduce health risks, including protecting pregnant women's unborn babies.

In 2015 the UK Committee on Carcinogenicity of Chemicals in Food, Consumer Products and the Environment (COC) reported on alcohol consumption and risks.[10] Its Lay Summary is stark: "Drinking alcohol has been

---

7. BBC, "1966: Aberfan." The 50th anniversary was remembered in 2016.

8. *Report of the Tribunal*, part VII, "Summary."

9. See Hore-Lacy, *Nuclear Energy*, 94. Table 15 has global energy-related accidents and deaths from 1977 to 2010.

10. COC is a group of independent experts advising UK government departments. and agencies.

shown to increase the *risk (or chance)* of getting some types of cancer. This does not mean that everyone who drinks alcohol will get cancer, but studies have shown that some cancers are more common in people who drink more alcohol."[11]

They investigated regular drinking habits (not binge drinking) for low, medium, and high drinking patterns with beers, wines, and spirits. For this they identified "statistically significant increased" cancer risks and concluded: "We found that 4–6% of all new cancers in the UK in 2013 were caused by alcohol consumption."[12]

Furthermore, the World Health Organization's 2010 "Global Strategy to Reduce the Harmful Use of Alcohol" identified alcohol as a significant contributor to global diseases and deaths (including road traffic accidents and collisions, violence, and suicide).[13] A 2016 fact sheet noted that annual worldwide deaths from misuse of alcohol total 3.3 million, which is 5.9 percent of all deaths, and there is a causal link with 200 diseases/injury conditions.[14] This is a massive worldwide health detriment, but alcohol remains socially acceptable.

What about tobacco smoking? Most people are aware that smoking is linked to cancer and serious health conditions. Cigarettes also contain radioactive material. The World Health Organization noted: "The tobacco epidemic is one of the biggest public health threats the world has ever faced, killing around 6 million people a year. More than 5 million of those deaths are the result of direct tobacco use while more than 600 000 are the result of non-smokers being exposed to second-hand smoke."[15] There are over four thousand chemicals in second-hand smoke and over fifty of these cause cancer. Yet, we are often given the impression that nuclear energy causes more harm than alcohol and cigarette smoking.

Discovery and Health

People worry about the risks from radiation, but before the nineteenth century we didn't know radioactivity existed. So without this knowledge there

---

11. COC, *Statement on Consumption*, ii. Italics original.
12. Ibid., iv.
13. WHO, "Global Strategy," 5.
14. WHO, "Alcohol." Alcohol also causes disability in addition to death.
15. Ibid.

was no fear. Then scientists discovered it. Radioactive radium was initially promoted as healthy.

However, studies began to show that high doses and repeated exposures produced risks of radiation-induced cancer, e.g., increased mortality among radiologists. In the US, women employed in dial-painting procedures used radium and pointed the brushes in their mouths. They ingested high levels of radium, which went to their bones and many died from cancer caused by high radiation doses.[16] Now radiation risks are recognized, but what about electricity itself? Is it dangerous?

Electricity Generation

For some people nothing is more suspect or devilish than nuclear energy. But is electricity a contender? A stupid question? But bear with me. In discussions on energy supplies it is an unspoken assumption that we must have electricity. Electricity is essential. Electricity production technology is taken for granted, whether it is from renewables, fossil fuels, hydro, or nuclear plants. It is widely treated as safe enough to have in our homes, schools, hospitals, workplaces, etc. Nevertheless, the development of electricity brought risks. For example, supplies to people's homes and factories initially had wires with no, or inadequate, safety insulation and no fuses or fuse boxes.[17] Electric shocks, fires, and fatalities were frequent.

However, we enjoy the benefits of electricity. But is it safe enough? In 2014 the UK's Electrical Safety First charity stated that annually in UK homes about 70 people die from electricity, and there are a staggering 350,000 serious electrical accidents. Moreover, 20,000 fires in homes are caused annually through electricity.[18] I recall sadly the fatal electrocution of a person putting up little Christmas tree lights. But let us agree that despite electricity's risks the option of going back to gas lights and candles is riskier—electricity is here to stay, provided we can generate it safely.

16. Martin et al., *Introduction to Radiation Protection*, 43.
17. For example, on the situation in the US, see Freeberg, *Age of Edison*, 201–5.
18. Electrical Safety First, "How Safe Is Your Home?," 4. Also see Electrical Safety First, "For DIYers," and their homepage for safety help and advice. Electrical Safety First, "How Safe Is Your Home?," 3, states: "Electrical Safety First is the UK charity dedicated to reducing deaths and injuries caused by electrical accidents. Our aim is to ensure everyone in the UK can use electricity safely."

Civil Nuclear Energy and Risks

At this stage its worth identifying the risks associated with electricity generation from nuclear plants. Environmentalist Stewart Brand stated:

> Older environmentalists talk about nuclear power exclusively in terms of what they see as the four great problems that condemn the technology–safety, cost, waste storage, and proliferation. These four have no forms of positive, only degrees of badness, and they are treated as absolutes. If a reactor accident is possible, then nuclear power is impossible; if the capital costs are high, then nuclear power is impossible, and so on. Absolutes are potent. Once something is seen as a capitalized Absolute Evil, it functions as a premise; everything has to exist in relation to your opposition to it.[19]

Other objections include: links with nuclear weapons, "too little, too late" (the claim that new nuclear plants cannot be build fast enough to reduce carbon emissions), unfulfilled dreams of electricity being "too cheap to meter," and insufficient uranium supplies.

Values, ethical issues, and moral concerns must, of course, be considered. Christian author Tim Cooper asked: "Nuclear Power—technological idolatry?" in a section on nuclear energy.[20] Then he added: "*No* nuclear reactor can be totally safe, because no technology is infallible and no human beings are perfect."[21] Later, in the 2006 Christian Ecology Link (CEL) policy paper against nuclear energy, Cooper stated: "The technology adopted in supplying energy should not involve excessive risk, including threats of pollution or warfare."[22] How should we respond?

Well, we must recognize that people can have genuine concerns on nuclear energy, but Brand (above) shows the negative views that abound. It is a negative nuclear narrative. This chapter begins to address such concerns (chapters 5 and 6 address specific issues more fully). There is, I suggest, an alternative, positive nuclear narrative.

In response to Cooper's question on technological idolatry, I should state that in my years as a scientist and tutor in the nuclear industry I was a monotheist. I didn't engage in technology worship, but true worship. And I wasn't alone. Of course, in this world nothing is absolutely safe. Even our

---

19. Brand, *Whole Earth Discipline*, 90.
20. Cooper, *Green Christianity* (1990), 200. Italics original.
21. Ibid., 202.
22. Cooper, *Faith and Power*, 5.

sun, which you recall from chapter 1 operates by nuclear fusion in God's created universe, doesn't provide total safety. Many people develop skin cancer from exposure to the sun's rays. Moreover, all forms of electricity generation have pros and cons that we must consider fairly in our options. Electricity itself is not totally safe, whatever means we use for its generation. Let us turn to electricity generation.

UK New Nuclear Plants: Justification for Building

The UK has a justification procedure to evaluate benefits against risks for new nuclear power stations. How has this happened? We will look at one example from the UK's Department of Energy and Climate Change (DECC). In 2010 the Secretary of State decided that ". . . *the AP1000 designed by Westinghouse Electric Company LLC*" is justified under *The Justification of Practices Involving Ionising Radiation Regulations 2004*.[23]

The Secretary noted the benefits:

- significant contribution to securing energy supplies,
- meeting low-carbon obligations,
- important economic benefits (e.g., global supply chain, improved workforce skills).

But this is followed by recognizing that the AP1000 will have a potential negative impact on health, safety, and the environment. This is considered to be small, well understood, and addressed by the current regulatory system.[24]

Exposures are "very small" compared with natural sources, but detriments are:

- radiation doses to the public, during normal operation, and accidents;
- radioactive waste and spent fuel arisings;
- environmental detriment, safety, and security.

---

23. DECC, *Justification of Practices*, 2. Italics original. DECC noted: "The AP1000 is a Pressurised Water Reactor (PWR), which is the most common type of nuclear reactor in operation throughout the world" (1). There are plans to build three AP1000 reactors at Moorside in Cumbria, UK. Each reactor will produce 1,135 MWe.

24. Ibid., 3, para. 1.7.

The Secretary of State declared that for the UK the benefits exceeded the detriments.[25] Nuclear power is considered to contribute to low-carbon sources, so that by 2050—together with renewables and fossil fuel plants using carbon capture and storage—"virtually all" UK electricity will come from these sources. Climate change is considered in the decision process.

In normal plant operation, the above report by DECC noted that the legal radiation dose limit from all sources (excluding natural sources and medical procedures) for members of the public is 1 mSv per year.[26] Moreover, it stated: ". . . a dose of 1 mSv per year is equivalent to an additional risk of fatal cancer of one in twenty thousand (0.005%) per year, and that a risk at this level is not detectable among normal background levels of cancer risk."[27] Cefas publishes annual reports on *Radioactivity in Food and the Environment* (see chapter 2 above) which demonstrate that public doses from civil nuclear plants are below the annual dose limit. Workers are also protected under legal dose limits and nuclear plants are highly regulated for safety and security.

Radiation Risks and Radon

One of the major concerns or objections to nuclear plants is the public's exposure to nuclear radiation from normal operations and accidents. But we need to see this in its context of our natural radiation background and artificial exposures. In chapter 2 we saw that 50 percent of the UK average annual individual dose from exposure to natural background comes from radon gas (1.3 mSv), while the total annual dose from natural and artificial exposure is 2.7 mSv. The World Health Organization commented:

> Radon is the second most important cause of lung cancer after smoking in many countries. Radon is much more likely to cause lung cancer in people who smoke, or who have smoked in the past, than in lifelong non-smokers. However, it is the primary cause of lung cancer among people who have never smoked.[28]

---

25. Ibid., 3, para. 1.9.

26. Ibid., 6, para. 1.24. In July 2016 DECC merged and became part of the Department for Business, Energy and Industrial Strategy (BEIS). The previous Department for Business, Innovation and Skills (BIS) is now part of BEIS.

27. Ibid., 6, para. 1.26. It draws upon the UK's Health Protection Agency (HPA) for this statement.

28. Zeeb and Shannoun, eds., *WHO Handbook on Indoor Radon*, 3.

RECOGNIZING RISKS AND BENEFITS

But what are the risks to people from radon gas?

Public Health England (PHE) compared radon deaths in the UK population with premature deaths from other causes. At around 1,100 deaths per year from lung cancer, the radon risk exceeds that of deaths for child cyclists, work accidents, and drunk driving.[29]

The average radon level in UK homes is 20 Bq/m^3. However, as the level rises in our homes so our risk rises. If someone lives in a house where the radon level is 200 Bq/m^3 then the level of risk exceeds, for example:

- "Radioactivity from fallout, nuclear accidents, waste disposal and nuclear power stations" or
- "Working in a nuclear power station."[30]

Besides radon exposure in homes, people are also exposed in schools and workplaces.

There is a higher risk for smokers and ex-smokers. Current smokers account for half of the 1,100 annual deaths. For example, the PHE table shows that at an indoor radon level of 20 Bq/m^3 the lifetime risk to a non-smoker is less than 1 in 200 but for a current smoker it is 1 in 7.[31] A 2009 Health Protection Agency report provides further information and recommendations for the UK.[32]

The World Health Organization offers a wider perspective on radon risks with an overview of major aspects impacting on national programs, i.e., planning, implementation, and evaluation. After a review of many scientific studies it concluded that radon exposure causes lung cancer and

---

29. PHE, "Radon at a Glance," under the heading "Radon risks." However, radon deaths are exceeded by road accidents (around 3,000 per year) and lung cancers in smokers (a staggering 28,000).

30. Ibid. As the indoor radon exposure level rises so does the risk.

31. PHE, "Risks to Your Health from Radon." Note that 1 Bq corresponds to one nuclear disintegration per second. It is the internationally recognized (SI) unit of radioactivity. Starting with the radioactivity expressed in becquerels, the dose in millisieverts (mSv) is derived. The earlier unit of radioactivity, the curie (Ci), is sometimes used, and then the dose is expressed in millirems (mrem). See the glossary.

32. HPA, *Radon and Public Health*. On simply reducing your risks, see PHE, "Risks to Your Health from Radon," with the four steps: (1) "Find out if you live in a radon risk area"; (2) "If you do, measure your home"; (3) "If the radon is high, reduce it"; (4) "If you smoke, give up."

globally the average indoor level of radon gas is estimated as 39 Bq/m^3, which is higher than the UK average.[33]

They show that the risk of radon-induced lung cancer depends upon whether the individual is a lifelong non-smoker, a current smoker, or an ex-smoker. Lifelong non-smokers have a lower risk than smokers. For example, WHO estimated that an indoor radon concentration of 0 or 100 Bq/m^3 carries a risk of fatal lung cancer (at age 75) of 4 or 5 in 1,000, respectively. Current smokers have much higher risks of 100 or 120 in 1,000, respectively. Ex-smokers have a much higher risk than lifelong non-smokers, but a much lower risk than current smokers.[34] So the choice to smoke coupled with natural exposure to indoor radon substantially raises an individual's risk of death from lung cancer.[35]

For example, the UK and US have a mean/average indoor concentration of 21 and 46 Bq/m^3, respectively, and the percentage of lung cancer arising from this radon exposure is 3.3 and 10–14 percent, respectively. This produced annual *estimated* numbers of radon-induced deaths from lung cancer of 1,089 (UK) and 15,400–21,800 (US).[36] And these are annual figures. The numbers are large. No wonder that natural radioactive radon is a major health issue requiring risk communication campaigns and prevention and mitigation strategies. Of course, knowing that something is a major health hazard and doing something about it by changing behavior is not straightforward, as David Hevey clearly discusses (see the quotation at the beginning of this chapter).[37]

But you may ask, "What has this to do with nuclear plants?" The WHO report has a chapter on "Radon Risk Communication." The first of five key messages stated the importance of communicating risk and messages on prevention, since ignorance of radon is widespread and the public may not see it as a health risk.[38] For WHO and national campaigns, a pri-

---

33. Zeeb and Shannoun, eds., *WHO Handbook on Indoor Radon*, 14.

34. Ibid., 16.

35. Note that smoking does not simply add to the risk of lung cancer from radon exposure, but it multiplies the risk by around 25 (ibid.). Also, see RPII and HSE, *Radon Gas in Ireland*, 2.

36. Zeeb and Shannoun, eds., *WHO Handbook on Indoor Radon*, 16. In Ireland there are up to 250 lung cancer cases each year attributed to radon gas. EPA (Ireland), *Radon and Your Health*.

37. Hevey, *Review of Public Information Programmes*, 14–15. Also, see Chow et al., *Evaluation and Equity Audit*.

38. Zeeb and Shannoun, eds., *WHO Handbook on Indoor Radon*, 73.

mary objective is clear and effective communication with the public and policy makers. The components are: assessing the public's perception of risk; delivering suitable risk messages; selecting target audiences; and possibly using comparisons, e.g., radon-induced lung cancer compared with other sources producing lung cancer. WHO has simple health messages to communicate but noted:

> ... that even in the context of professional health risk assessment, the term "risk" has many definitions. In general, a statement of risk to an individual requires a description of the probability or likelihood of harm and of the severity of the harm. In the case of radon, the harm is mainly lung cancer, which is a painful and fatal disease.[39]

Nevertheless, WHO found apathy among the public and policy makers, and a reluctance to be involved, even though radon is a major health hazard. People may object that it is natural and so okay, something we can't do anything about. But WHO rejects this argument as a misperception—high indoor radon concentrations are not entirely natural. We design and construct homes as residences. We have our own living habits. So radon is technologically enhanced and we can take action to prevent or remediate radon exposure.

Medical Radiation: Benefits and Risks

Examining medical radiation benefits and risks is a helpful way of putting into perspective radiation exposure from nuclear plants. The average person in the UK receives a small dose annually from medical exposure to radiation (see chapter 2). In the US the average annual population dose from medical exposures is higher than in the UK. Undoubtedly, the public recognizes the benefits of medical exposure for diagnosis and treatment. Nevertheless, such exposures may include fears over risks to patients.

Dauer and colleagues make this clear, realistically beginning their important article: "The mention of the word 'radiation' often evokes fear in patients, families, and health care professionals alike. Radiation is perceived as a unique hazard." They concede, however: "As public awareness of medical radiation exposure increases, there has been heightened awareness among patients and physicians of the importance of holistic

---

39. Ibid., 75–76.

benefit-and-risk discussions in shared medical decision making."[40] They recognized that public perception and fears arise from various sources, including: (1) radiation events causing injuries, (2) nuclear weapons/dirty bombs, (3) Chernobyl, and (4) comic books, e.g., Spider Man. The outcome "radiophobia" needs management, particularly of medical decisions regarding radiation. They are concerned that sensationalized media reports have "catalyzed public uneasy." Nevertheless, they admit that in the medical profession radiation overdoses have occurred, some examinations "may not be clinically justified," and that radiation doses received during diagnostic examinations are very variable. Together, these issues have triggered concerns "of the perceived seriousness of an accident or normal clinical practice."[41] Consequently, they recognize the challenge of convincing patients of the benefits and risks in a manner that is comprehensible, so that patients can give informed consent.

Balter and colleagues, in a clear 2011 article, considered the use of diagnostic imaging for patients and the associated radiation and nonradiation risks. They were concerned about responses to the US National Council on Radiation Protection and Measurements' 2009 publication on the increased US population doses from medical procedures (see chapter 2 above). In providing advice to clinical radiologists and referring physicians, they sensibly cautioned: "Too much attention to radiation is likely to distract awareness from the many—generally greater—nonradiation risks associated with most medical procedures."[42] Moreover, they warned that focusing on radiogenic risks (e.g., from CT scans) may distract from benefits to patients.

Medical CT Scans

In 2010, Jing Chen and Deborah Moir made the valuable point that significant societal benefits have been achieved through developments in medicine. Yet they wisely added: "Computed tomography (CT) is a powerful tool for accurate and effective diagnosis as it allows high-resolution three-dimensional images to be acquired very quickly. However, CT imaging tests, like other

40. Dauer et al., "Fears, Feelings, and Facts," 756. L. T. Dauer is at the Department of Medical Physics, Memorial Sloan-Kettering Cancer Center, New York.

41. Ibid.

42. Balter et al., "Radiation Is Not the Only Risk," 762. Balter and two of his coauthors are at Columbia University Medical Center, New York. The other is at Memorial Sloan-Kettering Cancer Centre, New York.

health care interventions, are not risk free."[43] Nevertheless, for most people the benefits outweigh the risks. People receiving CT scans are exposed to x-rays and associated risks. Their study estimates annual doses to the Canadian population from the rising numbers of CT scanners. Moreover, they helpfully compare individual CT doses with the natural doses from indoor radon gas and cosmic rays. Clearly, there are variations across regions but their study yields significant findings.

First, the typical effective dose averaged over the types of scan is 7.2 mSv.[44] In 2005–2006 there were 103 scans per 1,000 Canadians, thus giving an average dose of 0.74 mSv. When comparing the typical effective doses from CT scans over time, they found that the annual doses in:

- 1991 were below cosmic ray doses and just 17 percent of radon doses;
- 2006 were 2.4 times that from cosmic rays and 64 percent of radon doses.[45]

These are significant increases for individuals but they note these figures are *averages*. In any year some individuals may have no CT scans or none in a lifetime. Some individuals may receive multiple exposures. Thus a single scan to the abdomen or pelvis will give a dose of 10 mSv and one repeat scan in a year will give a dose of 20 mSv. Two repeats will give a total dose of 30 mSv.[46] As Chen and Moir observe, the Canadian radiation workers' annual dose limit is 20 mSv.

A study in 2015 examined multiple CT scans administered to individuals over a period of years. The authors assessed risks based on organ doses using recognized models. After five CT scans they showed that the risk exceeded a low risk band. Therefore: "Reported risk estimates demonstratively show that repeated CT exposures for diagnostic purposes should be minimized; they can be recommended for prescription if it is clinically justified and medically necessary."[47]

43. Chen and Moir, "Estimation of Annual Effective Dose," 131. When they wrote their article, Chen and Moir worked in the Radiation Protection Bureau, Health Canada.

44. Ibid., 133. Abdomen/pelvis and spine scans give the highest doses of 10 mSv, while a typical dose to the brain is 2 mSv.

45. Ibid., 135. The annual doses from cosmic rays and radon gas are 0.31 mSv and 1.15 mSv, respectively.

46. Ibid., 135–36.

47. Ivanov et al., "Estimating the Lifetime Risk," 840. V. K. Ivanov is from the Federal State Institution Medical Radiological Research Centre of the Russian Ministry of Health, Obninsk.

Finally, an important 2016 article by Sacks and colleagues challenged the current method of calculating exposure risks from exposure to low levels of radiation. It had produced fear. This method has led to "mass radiophobia," unnecessary evacuations at Fukushima (Japan), patients refusing medical imaging studies, and the common unwarranted dislike of nuclear energy.[48] In particular they noted: "This often-reflexive fear reaction inhibits many from even seriously investigating the use of nuclear energy."[49,50]

Electricity and Health: Benefits and Risks

Air Pollution, Illness, and Deaths

In the medical journal *The Lancet*, Markandya and Wilkinson wrote an important 2007 article, "Electricity Generation and Health." From an historical survey, they noted electricity's generally positive role in enhancing well-being, but recognized the arrival of new problems: "The burning of large amounts of fossil fuels to produce the electricity we demand generates emissions that are harmful to health and are a source of climate change."[51] They addressed health effects from electricity generation in both developed and developing countries.

What did they discover for Europe? The category of impact for both mortality and morbidity was identified for the pollutant and then the effects, e.g., life reduction, cancers. They examined six primary sources of electricity generation (lignite, coal, gas, oil, biomass, and nuclear) and compared these under: "deaths from accidents" (public and occupational persons) and "air pollution–related effects" (deaths, serious illness, and minor illness). The

---

48. Sacks et al., "Epidemiology Without Biology," 20. Their comments do not only apply to Japan. They reckon that the linear no-threshold (LNT) hypothesis used to determine risk actually overstates the risk, and they are not alone here. See Allison, *Nuclear Is for Life* (2015), 11–12. Wade Allison is Emeritus Professor of Physics, Fellow of Keble College, University of Oxford.

49. Sacks et al., "Epidemiology Without Biology," 22.

50. For further discussions on medical issues and medical x-rays with associated risks, see Elliott, "Issues in Medical Exposures"; Wall et al., "Radiation Risks."

51. Markandya and Wilkinson, "Electricity Generation and Health," 980. Professor Anil Markandya, Department of Economics and International Development, University of Bath, and Paul Wilkinson, London School of Hygiene and Tropical Medicine.

outcomes were summarized in Table 2, "Health effects of electricity generation in Europe by primary energy source (deaths/cases per TWh)."[52]

The health burden from nuclear generated electricity was much lower than from fossil fuels, with gas in one category of occupational deaths lower than nuclear. Serious health problems among coal miners were recorded, with up to 12 percent having "potentially fatal diseases."[53] Their figure 2 compared the eight primary energy sources and showed that nuclear had the lowest direct and indirect emissions.[54]

Nonetheless, while recognizing the considerable health benefits of current nuclear plants, the authors raised issues relating to accidents, waste, and approval/construction times. Nuclear fusion they considered a long-term benefit, as it is not currently commercially available. They warned that failing to replace the 17-percent nuclear contribution to global electricity production now will add a substantial barrier to urgent reductions of greenhouse gases. In conclusion, they commented that nuclear's future energy role "depends on a balance of (perceived) risks."[55]

Similar studies have demonstrated the health benefits of nuclear energy, so let's look at them. Pushker Kharecha and James Hansen have written valuable articles, adding to Markandya and Wilkinson's earlier work, e.g., their 2013 paper, "Prevented Mortality and Greenhouse Gas Emissions from Historical and Projected Nuclear Power."[56]

The environmentalist Mark Lynas helpfully drew attention to this benefit: "In a landmark 2013 paper, climatologists James Hansen and Pusher Kharecha calculated that the use of nuclear power between 1971 and 2009 avoided the premature deaths of 1.84 million people thanks to its air pollution benefits."[57] Truly, this is a massive number of lives saved. And Kharecha and Hansen's article took account of the three major nuclear accidents. Yes,

---

52. Ibid., 981. Note that the unit is TWh (terawatt hour), which means the energy sources are compared against the same amount of electricity generated. No figures were provided for deaths and accidents from biomass and no occupational deaths were given for oil. Nuclear was the only energy source without minor illnesses.

53. Ibid.

54. Ibid., 982. Fig. 2 is "Full Energy Chain CO2 Equivalent Emissions by Primary Energy Source."

55. Ibid., 988.

56. At that time, they were both working at the NASA Goddard Institute for Space Studies and Columbia University Earth Institute, New York.

57. Lynas, *Nuclear 2.0*, 56.

their article was challenged by some authors, but they strongly defended their analysis.[58]

Indeed, in another article, "Coal and Gas Are Far More Harmful than Nuclear Power," they argued that both anthropogenic climate change and air pollution from burning fossil fuel can be mitigated through concurrently employing nuclear power, renewables, and greater efficiency. Again, they explain: "nuclear power prevented an average of over 1.8 million net deaths worldwide between 1971–2009."[59] Finally, in another valuable 2013 article Hansen and colleagues examined climate change, young people, future generations, and nature. They concluded that replacing fossil fuels was a huge task and, as above, that nuclear energy should be part of the mix, which should be determined locally.[60]

Additionally, Qvist and Brook's 2015 article addressed Sweden possibly phasing out its nuclear plants and the resultant impacts on health, the environment, and economy.[61] They informed us that major political parties (including the Green Party in the coalition government) wished to prematurely decommission a nuclear plant. Indeed, the political decision to prematurely shut down the Barsebäck plant has resulted in an estimated "2400 avoidable energy-production-related deaths and an increase in global CO_2 emissions of 95 million tonnes" as of October 2014.[62]

In a persuasive case, Qvist and Brook argued against this phase-out policy, highlighting that:

- Sweden's reactor fleet is only just over halfway through its productive life;
- forced shut down would produce up to 2.1 Gt of additional CO_2 emissions;
- 50,000–60,000 energy-related deaths could be prevented by continued operation;
- nuclear phase-out would mean a retrograde step for climate, health, and economy.[63]

---

58. See Kharecha and Hansen, "Response to Comment by Rabilloud" and "Response to Comment on." Their second response was made to Sovacool et al.

59. Kharecha and Hansen, "Coal and Gas."

60. Hansen et al., "Assessing 'Dangerous Climate Change.'"

61. Qvist and Brook, "Environmental and Health Impacts," 1.

62. Ibid.

63. Ibid. Note: 1 Gt (gigatonne) is 1,000 Mt (megatonnes).

The estimated loss is 100 billion US dollars. Qvist and Brook's analysis assumed three scenarios, in which nuclear was replaced by: (1) 100 percent coal, (2) 90 percent coal with 10 percent natural gas, or (3) 80 percent coal with 20 percent natural gas.

Although Sweden has many dams, its hydro capacity cannot be increased substantially. A commonly held belief is that severe nuclear accidents are more serious than dam failures. Qvist and Brook assessed the risks of dam failures and consequences, however, and showed that this popular belief is incorrect. They noted that in the history of Swedish nuclear plants there have been no accidents impacting people or the environment. Yet Sweden had experienced several dam failures, resulting in one death and damaged roads and bridges. The worse dam disaster occurred in China (1975) at the Banqiao Reservoir Dam (and dams further downstream). Qvist and Brook called this "the most severe energy-production-related accident in history."[64] Around a staggering 230,000 people died, while 6 million buildings collapsed affecting 11 million residents. The article compared the potential impacts of failures in Sweden's dams with nuclear accidents and showed how many more people could be killed and cities destroyed through dam failures.[65]

In June 2016, a Swedish political framework agreement specified that a current nuclear capacity tax would be phased out over a two-year period (thus removing an unfair tax), and as nuclear plants retired new nuclear plants (up to ten) could be constructed on existing nuclear sites. It was seen as a positive step by the World Nuclear Association.[66]

Historical Prevention of Carbon Dioxide Emissions through Nuclear Generation

Another benefit of nuclear power, besides energy security, is frequently overlooked: its role historically in limiting carbon dioxide emissions. Nuclear has generated clean electricity for decades. Kharecha and Hansen calculate that nuclear plants prevented around 64 gigatons (Gt) of CO_2-equivalent emissions from greenhouse gases in the period 1971–2009. Putting this in perspective:

64. Ibid., 6.
65. Ibid., 6–7.
66. *WNN*, "Sweden Abolishes Nuclear Tax." For information on Sweden, see WNA, "Nuclear Power in Sweden."

> This is about 15 times more emissions than it caused. It is equivalent to the past 35 years of CO_2 emissions from coal burning in the U.S. or 17 years in China . . .—i.e., historical nuclear energy production has prevented the building of hundreds of large coal-fired power plants.[67]

This is a substantial saving in damaging emissions and a benefit to the whole planet. Without this valuable contribution the world would be further down the undesirable pathway of dangerous climate change.

Power and Nonproliferation: Megatons to Megawatts

Another concern about nuclear power is the possibility of proliferation, when nuclear fuel can be turned into nuclear weapons, e.g., using the enriched uranium or plutonium formed within the fuel. Moreover, Britain, for example, has substantial stocks of plutonium that need to be guarded to prevent it falling into the hands of terrorists.

Mark Lynas recognizes this real concern but responds: "Nuclear is a classic example of a dual-use technology, but banning it altogether makes no more sense than trying to ban agriculturally vital nitrate fertilisers because they can be used to manufacture terrorist car bombs."[68] He continues: "You have to beat swords into ploughshares, not try to ban them both."[69] Stewart Brand is crystal clear: "it should be said that nuclear energy has done more to eliminate existing nuclear weapons from the world than any other activity."[70] This seems to be an absurd remark until we realize that he refers to the US and Russian nonproliferation program informally called Megatons to Megawatts. In this program, which ran from 1994 to 2013, Russian uranium from weapons was turned into fuel for US nuclear reactors to produce energy. Over 15 years it generated 10 percent of all US electricity (or around half of all nuclear power production)[71]—truly, a large contribution to nonproliferation and clean energy.

In conclusion, the benefits of nuclear power on human health and environment are considerable. Nuclear generation has substantially reduced

---

67. Kharecha and Hansen, "Coal and Gas."
68. Lynas, *Nuclear 2.0*, 64.
69. Ibid., 65.
70. Bland, *Whole Earth Discipline*, 108.
71. Stover, "More Megatons to Megawatts." Also, see Bawden, "How America Uses Russian Warheads," 12.

premature deaths, serious illnesses, and minor illnesses. Historically worldwide, considerable carbon dioxide discharges have been prevented which would have driven us further towards increased climate change. Continued use of nuclear further reduces these damaging discharges. Moreover, nuclear gives us a baseload secure energy supply.[72] As seen in the Megatons to Megawatts program, nuclear power can aid in nonproliferation and produce clean electricity. It contributes to job creation and affordable, reliable energy supplies. Its electricity can be used in changing transport from fossil fuels to electrical vehicles. Nonetheless, fear of nuclear power and radiation often seems to outweigh these many benefits.

Nuclear Power: Professional and Public Radiation Risk Perceptions

In 2013 Tanja Perko wrote the essay "Radiation Risk Perception: A Discrepancy between the Experts and the General Population." The study examined the perception of lay people and experts regarding radiological risks ("nuclear waste, medical x-rays, natural radiation, an accident at a nuclear installation in general, and the Fukushima accident in particular"). A link was made between risk perception and risk communication by using media presentations of the risks. The analysis concluded (shown in other studies) that experts and the general public have different perceptions of radiological risk. There is a wide communications gap. Moreover, mass media's language is different from that of technical experts.[73]

Other studies have shown similar issues, and not just in radiological risks related to nuclear power plants. For example, Stewart and colleagues described ten environmental health case studies in an industrial area of North West England. Their focus was on risk perceptions by the public regarding environmental health issues, including "waste incineration, land contamination, odour and air contamination, non-ionising radiation, acute chemical incidents, flooding and cancer due to environmental risks."[74] Although power generation was excluded the report gave valuable

---

72. The EIA's "Glossary" defines "baseload" as: "The minimum amount of electric power delivered or required over a given period of time at a steady rate."

73. Perko, "Radiation Risk Perception," abstract. Perko is from the Belgian Nuclear Research Centre SCK CEN.

74. Stewart et al., "Real or Illusory?," 1159. The authors are from the Cheshire & Merseyside Health Protection Unit, Kirby, Merseyside, and the Centre for Public Health,

insights into public perceptions of risks in non-nuclear industries. It is not just nuclear technology that causes concern among members of the public.

Radioactive Waste and Low Doses

In 2009, Mobbs and colleagues from the UK's Health Protection Agency (HPA), published the report *An Introduction to the Estimation of Risks Arising from Exposure to Low Doses of Ionising Radiation*. They noted that the main effects of low doses are a small increase in the risk of cancer (which is proportional to the dose received). Nevertheless, "risk is a random process," they noted, and just because someone is exposed to radiation doesn't mean that person will get cancer in the future. If doses are low to a small population, then risks are likewise low and no one may develop cancer from the exposure.[75] Their study particularly addresses radiation exposure from radioactive waste and gives two helpful examples to put doses in context.

First, consider a worker who received a dose of 20 μSv per year from handling Very Low Level radioactive waste.[76] The fatal cancer risk specified by the International Commission on Radiation Protection was used. This gives, for 20 μSv per year, a risk of one in a million, a tiny percentage of 0.0001. But what does this tell us? Well, the UK average risk of dying from cancer is 25 percent. So for each worker exposed to this level (20 μSv per year) the risk rises from 25 percent to just 25.0001 percent—an increase too small to be observed in the population. In other words: "this indicates that if 1 million workers handled this waste and each received this dose, then only one would be expected to die of a corresponding radiation induced cancer while about 250000 naturally occurring cancers would be expected."[77] Nonetheless, they observe two reasons why this one cancer could not be actually observed: (1) natural cancer levels vary in the population, and (2) radiation-induced cancers cannot be distinguished from other cancers.

The second example is based on the UK dose limits for members of the public: 1 mSv per year. The additional fatal cancer risk to somebody exposed

---

Research Directorate, Faculty of Health and Applied Social Sciences, Liverpool John Moores University.

    75. Mobbs et al., *Introduction to the Estimation of Risks*, 5–6.
    76. The units here are in μSv (microsieverts). As a reminder, 1,000 μSv = 1 mSv.
    77. Mobbs et al., *Introduction to the Estimation of Risks*, 6.

to this dose limit is "one in twenty thousand (0.005%) per year."[78] This also would not be detectable above the background levels of cancer risk.

Let us compare these risks with our discussion in chapter 2 on the levels of natural radiation background and artificial sources. The UK's average annual dose is 2.2 mSv from natural sources and if we add artificial sources the total is 2.7 mSv. The figure of 20 µSv per year (used above) is just 0.75 percent of 2.7 mSv.[79] Moreover, 20 µSv is the dose someone receives by eating just one bag of 100 g of Brazil nuts twice in a year!

It is true that there are some "unavoidable uncertainties" in the risk estimates used. Moreover, some people have challenged ICRP's risk factors but the HPA, from extensive evidence collected and many reviews, concluded it "has confidence that the radiation risk factors used by ICRP provide a sound basis for a radiological protection system."[80]

Morally Responsible Risk Communication

Finally, a "three-level framework" has been developed for morally responsible risk communication on nuclear risks in a 2016 paper. The authors examined the procedure, message, and the effects of communication.[81] It is not just about communicating numbers, as they recognized the role of emotions (e.g., sympathy) and values. Their suggested model noted: "A successful risk communication effort triggers reflection, compassion and a willingness to take responsibility for energy-related issues."[82] Such an approach has merits that will appeal to Christians.

Conclusion

Nuclear energy must be seen within the context of the benefits and risks posed by other forms of electricity generation alongside other risks in life. It provides a clean low-carbon energy source, security of supply, and affordable energy. Further benefits include the Megatons to Megawatts program

78. Ibid.
79. Ibid., 7.
80. Ibid., 13.
81. Fahlquist and Roeser, "Nuclear Energy, Responsible Risk Communication," abstract. They are in the Philosophy Section, Faculty of Technology, Policy and Management, Delft University of Technology, Netherlands.
82. Ibid.

and job creation. The public doses received in normal operation are small compared to natural and artificial sources of radiation and exposure to radon gas in our homes. Beware of radiophobia without facts.

From a Christian perspective, being concerned about our world means assessing the benefits and risks of different technologies; it is essential and an integral part of Christian mission. New nuclear plants should be advocated as providing a balanced energy mix to avert dangerous climate change. We must seek God's guidance for the challenges ahead. Next we examine policies and positions on nuclear energy.

Further Reading

Fischhoff, Baruch, and John David Kadvany. *Risk: A Very Short Introduction*. Oxford: Oxford University Press, 2011.
Public Health England. "Radon at a Glance." See the section "Radon risks." http://www.ukradon.org/information/radonataglance.
World Health Organization. "Alcohol." Fact sheet, January 2015. http://www.who.int/mediacentre/factsheets/fs349/en/#.
———. *Communicating Radiation Risks in Paediatric Imaging: Information to Support Healthcare Discussions about Benefit and Risk*. 2016. http://www.who.int/ionizing_radiation/pub_meet/radiation-risks-paediatric-imaging/en/.
———. "Ionizing Radiation, Health Effects and Protective Measures." Fact sheet, April 2016. http://www.who.int/mediacentre/factsheets/fs371/en/.
———. "Tobacco." Fact sheet, June 2016. http://www.who.int/mediacentre/factsheets/fs339/en/.

Questions

1. What are your views on the cancer risks associated with alcohol, tobacco, and radon gas in homes? Are these risks widely known about or understood by the general public?

2. Read a copy of environmentalist George Monbiot's 2013 newspaper article "Power Crazed" (http://www.monbiot.com/2013/12/16/power-crazed/). He asked: "Why do we transfer the real health risks inflicted by coal onto nuclear energy?" Can you summarize the various risk comparisons he makes? How does his approach influence your views on nuclear energy?

3. How has this chapter changed, or challenged, your views on nuclear energy?

4. Philip Thomas is Professor of Risk Management at the University of Bristol, UK. Read his 2016 article "Why We Need a New Science of Safety" (http://theconversation.com/why-we-need-a-new-science-of-safety-58436). How does he help us think about safety and risk in our daily lives?

4

Energy Policies and Positions

Since nuclear power generation does not emit significant quantities of the acid gases and, in the UK gas-cooled reactors, a very much smaller amount of CO_2, it has played a significant role to date in reducing emissions of these gases in the UK.

CHURCH OF SCOTLAND[1]

More than any other single issue, it was climate change that brought me into politics, and it is climate change that keeps me awake at night.

CAROLINE LUCAS[2]

Use of nuclear energy has avoided the release of 56 Gt of CO_2 since 1971, equivalent to almost two years of global emissions at current rates.

INTERNATIONAL ENERGY AGENCY[3]

---

1. SRTP, "What Future for Nuclear Power?"

2. Lucas, *Honourable Friends?*, 103. Caroline Lucas was elected as Britain's first Green member of parliament (MP) on May 7, 2010.

3. IEA, *Energy and Climate Change* (2015), 73. The report notes nuclear has an important role in curbing growth of greenhouse emissions in what it terms the Bridge Scenario up to 2020 (box 3.2).

Introduction

THIS CHAPTER IS MAINLY descriptive, outlining energy policies and positions, primarily in the UK. For readers who dislike many numbers, please note that after the section "Electrical Generation Capacity" I avoid using many numbers. Admittedly, this is a long chapter, but some sections may be omitted without losing the main thrust. Public attitudes to nuclear energy are addressed in chapter 5.

People and politicians often agree on rapidly reducing our fossil fuel dependency, but often disagree on ways forward. Sometimes acrimonious arguments polarize people. My view is that we need nuclear and renewables to avoid adverse climate change, and I agree with Dawn Stover that they can complement each other.[4] So let us start our journey.

Electricity Generation: The Role of Nuclear Energy

The International Energy Agency (IEA) is an authoritative source on energy.[5] Its 2009 *Ensuring Green Growth in a Time of Economic Crisis* stated: "It is important to note that *all* options are needed. It is not either CCS [carbon capture and storage] or renewables, nor nuclear or energy efficiency, but all of them together."[6] This is the position I adopt. However, as my focus is on nuclear energy I will look at figures on the number of operable reactors worldwide, those being constructed and those being closed down (decommissioned). The World Nuclear Association reported on civil reactors for 2015:

- 436 operable reactors worldwide (beginning of 2015);
- 439 operable reactors worldwide (end of 2015)[7];

4. Stover, "Nuclear vs. Renewables."
5. The IEA is an autonomous agency (founded 1974) with the primary mandate: "to promote energy security amongst its member countries through collective response to physical disruptions in oil supply, and provide authoritative research and analysis on ways to ensure reliable, affordable and clean energy for its 29 member countries and beyond." IEA, *Tracking Clean Energy Progress 2016*, 2.
6. IEA, *Ensuring Green Growth*, 6. Italics original.
7. Operable reactors include operating reactors and those that are operable but not currently operating for generating electricity. In Japan, following Fukushima, all the reactors were closed for safety checks. Reactors are restarting and others are operable but going through the process of restarting. Some were permanently closed. Just quoting numbers of "operating reactors" without clarification overlooks the Japanese reactors that have passed safety checks and are now operable but awaiting a restart.

- more reactors under construction now than in any time in the previous 25 years;
- net increase in generation capacity is 9,875 MW[8];
- 2,441 TWh generated (about 10 percent of worldwide electricity generation; one-third of low-carbon electricity production);
- average capacity factor globally (excluding Japan) almost 82 percent;
- 10 new reactors came online;
- for reactors starting in 2015 the average construction time was 73 months;
- 66 civil reactors under construction at end of 2015 with 158 more planned.[9]

Reactors have a finite life so in 2015 seven reactors worldwide were permanently closed down for decommissioning.[10] The IEA notes for 2015 the news of the restart of Japan's first two reactors after all the nation's reactors were closed following the Fukushima accident.[11] China is the world leader in construction of new nuclear plants. In the last three years, the IEA notes China has connected 14 reactors to the electricity grid and their average construction time was 5.5 years.[12]

Nevertheless, the World Nuclear Association realistically recognizes challenges facing the nuclear industry: "The rate of new build is, however, insufficient if the world is to meet the targets for reducing the impacts of global warming agreed at the 21st Conference of Parties (COP21) on

---

8. MW here refers to megawatts of electricity, usually written MWe.

9. WNA, *World Nuclear Performance Report 2016*, 2, 10–13, 16. Note that "The average capacity factor, which reflects the actual amount of electricity provided to the grid as a percentage of the maximum possible, has been over 80% since the start of the century" (10). This is a high capacity and greater than that for wind or solar energy. However, nuclear plants globally require, for example, refueling and routine maintenance. So, they do not achieve 100 percent.

10. Ibid., 21. Germany's one reactor closed for political reasons and Japan's five reactors were shut down after the 2011 Fukushima accident and the decision was taken not to restart them.

11. Sendai 1 and Sendai 2 restarted commercial operation in 2015. In September 2016, the Ikata nuclear power plant Unit 3 in the Ehime prefecture reentered commercial operation. Twenty more reactors are moving towards restarting. *WNN*, "Ikata 3 Back." However, Ikata 1 will be decommissioned. *WNN*, "Shikoku Decides."

12. IEA, *Tracking Clean Energy Progress 2016*, 24. See also IEA, *Energy and Climate Change*; and WEC, *World Energy Issues Monitor 2016*.

climate change, which took place in Paris last year."[13] And here the IEA's assessment agrees: "Long-term policy and financial uncertainty remain for nuclear power, but significant increases in both construction starts and grid connections in 2015 helped make progress towards meeting 2DS targets."[14]

In January 2017, the World Nuclear Association issued new figures on nuclear capacity.[15] At the end of 2016, worldwide there were 447 operable reactors (providing 391.4 GWe net) with another 60 reactors being constructed (which will provide 64.5 GWe gross). This compares with 439 operable reactors (382.2 GWe) at the end of 2015. Note my discussion in the footnote above on "operable" reactors, as this unqualified term is potentially misleading. Japanese reactors awaiting restart are considered "operable." However, progress is being made here and the US Nuclear Energy Institute reported in March 2017 that 26 reactors at 16 Japanese sites have applied to the Nuclear Regulatory Authority to restart.[16]

So, what is the situation on the UK's electricity supply from its reactors? Well, in 2015 nuclear energy supplied 19 percent of the UK's electricity (up from 17 percent in 2014).[17] Moreover, in 2015 the total amount of energy obtained from low-carbon sources was 16.5 percent, with nuclear energy contribution at 48 percent of low-carbon sources.[18] The UK's nine operating civil nuclear power plants owned and operated by EDF Energy are all AGRs (advanced gas-cooled reactors) except for one PWR (pressurized water reactor).[19] In February 2016, EDF reported life extensions for these AGRs beyond the scheduled closure dates to help maintain Britain's electricity supply.[20] On August 2, 2016 Heysham 2 nuclear plant Unit 2

13. WNA, *World Nuclear Performance Report 2016*, 2. Comment by Agneta Rising, Director General of the WNA.

14. IEA, *Tracking Clean Energy Progress 2016*, 11. Here 2DS means 2 °C scenario.

15. Information from *WNN*, "Worldwide Nuclear Capacity."

16. NEI, "Japan Nuclear Update: NRA Considering Restart." For more information on restart estimates by the Institute of Energy Economics, Japan, see the July 2016 summary, *WNN*, "Japanese Institute Sees 19 Reactor Restarts."

17. BEIS, *UK Energy in Brief 2016*, 25.

18. Ibid., 11. The total amount of energy is called the primary energy as distinct from electricity generated.

19. The AGRs, commissioned in the 1970s to 1980s, are: Hinkley Point B, Hunterston, Hartlepool, Heysham 1, Dungeness B, Heysham 2, and Torness. Each AGR site has two reactors, once owned by British Energy but purchased by EDF (2008). The PWR site is Sizewell B. All Magnox plants are now being decommissioned.

20. *WNN*, "EDF Energy Extends." Hartlepool and Heysham 1 (by five years to 2024); Heysham 2 and Torness (by seven years to 2030). These plants alone supply electricity

achieved the world record for continuous operation of a commercial nuclear reactor at 895 days.[21] It continued operation until its scheduled maintenance/inspection outage (when the reactor is temporarily shut down and electricity is not being generated) on September 16, 2016, by which time it had increased its world record to a 940-day run.[22]

Furthermore, *World Nuclear News* reported two important issues. First, EDF reckoned that these AGR total life extensions potentially avoid "80 million tonnes of carbon dioxide emissions, equivalent to taking all the cars off the road in the UK for three-and-a-half years."[23] This shows nuclear's contribution to clean energy. Second, in 2015 EDF's UK nuclear reactors together generated electricity for 60.6 terawatt hours and achieved "the highest level for ten years and 50% higher than in 2008 when the company acquired the plants from British Energy."[24]

Moreover, the UK's Committee on Climate Change (CCC) comments on harmful emissions:

> Power sector emissions were 121 $MtCO_2$ in 2014, around a quarter of total UK greenhouse gases.... The majority of these emissions are from coal (71% of emissions, whilst providing 32% of generation) followed by gas (27% of emissions, 29% of generation). There are no direct emissions from the 19% of generation from nuclear or 20% of generation from renewables.[25]

Note that nuclear and renewables have no harmful emissions. The CCC recognizes the current role of nuclear energy, but most current nuclear plants will close by 2030. Moreover, there will be closures of coal plants, but increased electricity demand (from more electric vehicles and heat pumps) will be partly offset by improved energy efficiency. The overall assessment from plant

---

to around one quarter of all UK homes. Life extensions previously announced for EDF's other AGR stations were: Hinkley Point B and Hunterston B (to close in 2023) and Dungeness B (to close 2028). Sizewell B's scheduled closure date is 2035.

21. *WNN*, "British Reactor Takes Record." Heysham 2, on the northwest coast of England, has two reactors. I had the privilege of going inside one reactor during site construction.

22. *WNN*, "Record 940 Days of Continuous Operation." In this period, it generated over 14 TWh electricity and avoided carbon dioxide emissions of more than 7 million tons.

23. *WNN*, "EDF Energy Extends."

24. Ibid.

25. CCC, *Power Sector Scenarios*, 9. The Committee on Climate Change is an independent, public body under the Climate Change Act 2008. It advises the UK government, Parliament, etc., on reducing emissions and preparations for change.

closures and increased demand for the 2020s means the UK will need new generation and capacity.[26] Consequently, the CCC considers a role for future nuclear even with delays and the contract price of Hinkley Point C. But it expresses concerns on escalating costs which could question a nuclear program.[27]

The UK's National Grid's *Future Energy Scenarios in Five Minutes* for 2016 identifies the three key technologies for electricity decarbonization as: nuclear, renewables, and carbon capture and storage (CCS) engaged generation. Its message is clear: "The cost-optimal pathway utilises all three of these technologies; approximately 22 GW of nuclear, 100 GW of renewables and 20 GW of CCS in 2050."[28] Electricity storage and interconnection to other countries will increase.

Note that electricity consumption is a part of all energy consumption. For example, natural gas is used for heating in homes and transport uses fuel. It is expected that world energy consumption will rise almost 50 percent from 2012 to 2040, while electricity generation increases by 69 percent.[29] In this time frame the fastest growing sources for electricity will be renewables, natural gas, and then nuclear. Moreover, in the Organisation for Economic Co-operation and Development (OECD) countries, by 2040 there is a projected drop overall in nuclear.

It's true that Germany has closed down some nuclear plants and will phase out its remainder, but this has adversely affected greenhouse gas (GHG) emissions. Moreover, German utilities have brought court action against the government for the early phase-out of their nuclear reactors and it looks like the government will pay billions of euros in compensation.[30]

---

26. Ibid. Up to 200 TWh per year and at least 20 GWe.

27. Ibid., 13. However, its future scenarios investigate new nuclear together with renewables and carbon storage and capture (CCS). See, for example, ibid., 17. Scenarios without new nuclear and CCS are also considered.

28. National Grid, *Future Energy Scenarios in Five Minutes*, 10. It states action is required this decade to head for the UK's 2050 carbon reduction target (emissions reduced 80 percent from 1990 levels). The National Grid's *Future Energy Scenarios* (full report) notes two at least of the three key technologies (nuclear, renewables, and CCS) are required, but the cost-optimal pathway utilizes all three technologies and meets the 2050 target on time (p. 9). With only two technologies costs increase substantially (pp. 159, 165).

29. *WNN*, "EIA Sees Strong Growth." EIA is the Energy Information Administration, US Department of Energy.

30. *WNN*, "Court Backs German Utilities." *WNN* stated: "Germany's Federal Constitutional Court today ruled that, although the country's 2011 phase-out legislation is essentially in compliance with the constitution, power utilities are entitled to 'reasonable' compensation for the early shut down of their nuclear power reactors."

Furthermore, in Belgium the phasing out of nuclear has been seriously questioned in the International Energy Agency (IEA) 2016 report. It notes that long-term temporary nuclear plant closures recently reduced nuclear's contribution to electricity generation; Belgium's nuclear power share is nevertheless among the highest in OECD countries.[31] Yet changing circumstances there meant: "The law on the nuclear phase-out of 2003 was changed in 2015 to enable the long-term operation of the two oldest nuclear reactors ... by ten years until 2025."[32] However, the IEA warns:

> The current policy to phase out all nuclear power plants by 2025 ... does not help Belgium meet any of its energy policy goals. In contrast, it adds to the generation adequacy problem, increases CO_2 emissions and increases the costs of generating electricity. The government should consider whether the current phase-out policy is optimal for securing affordable low-carbon electricity for Belgium.[33]

Nuclear has also performed an important role in reducing air pollution. Nevertheless, security of electricity supply and concerns on phase-out costs means the Belgian government has been advised to rethink its current phase-out policy in the light of the better option of allowing nuclear power plants: "to run as long as the regulator considers them safe. The IEA recommends the government to simply avoid a phase-out as it is currently envisaged."[34]

Elsewhere, new nuclear plants are being built. In the US, Tennessee Valley Authority (TVA) started to build its Watts Bar 2 reactor in 1972, but this was suspended in 1985. Work resumed in 2007 and in May 2016 it achieved criticality, the first US reactor to start up since 1996, when Watts Bar 1 started.[35] On October 19, 2016 Watts Bar 2 began commercial operation.[36]

Moreover, a Senate Energy and Natural Resources Committee hearing (May 17, 2016) on advanced reactor technologies heard witnesses supporting advanced reactors, international prospects, and global leadership

---

31. IEA, *Energy Policies . . . Belgium*, 129. See ch. 10, "Nuclear Energy," for further information. For a summary of the report, see NucNet, "Belgium Closures."

32. IEA, *Energy Policies . . . Belgium*, 112.

33. Ibid., 114.

34. Ibid., 142.

35. *WNN*, "First Criticality." It includes upgrades from lessons learnt from the 2011 Fukushima accident.

36. *WNN*, "Watts Bar 2 Begins." Watts Bar is near Spring City, Tennessee. It was finished on budget.

opportunities.³⁷ Of course, there are still issues in the US with nuclear power plants closing prematurely, as a May 2016 Department of Energy (DOE) summit demonstrates.³⁸ What then of the new Trump administration? In the UK, Christian author Martin Hodson's paper *Donald Trump, the Environment and the Church* (January 2017) rightly raising concerns from a Christian perspective over future environment policies, energy, and climate change. In his conclusion, he leaves us with ten key questions to be answered in the next four years.³⁹

I do not wish to repeat his helpful analysis but, rather, I do wish to draw attention to important articles on nuclear energy in the *Bulletin of Atomic Scientists* (2017) that were written particularly to accurately inform the Trump administration on the range of views about nuclear energy. John Mecklin (editor-in-chief) introduced leading experts' views on whether nuclear energy can make significant contributions to addressing climate change concerns and greenhouse gas emissions. He sensibly states that the *Bulletin* "provides research and analysis that should be taken into account, before hard decisions on climate change and energy policy are made by the United States government, or any other."⁴⁰ I particularly found Dawn Stover's interview with the climate scientist Kerry Emanuel very valuable, as he persuasively presented the case for more nuclear energy in combatting climate change.⁴¹ It remains to be seen how the Trump administration moves forward from here. There is no reason for complacency if we want clean air—many choices and challenges remain.

I hope examples above, and below, demonstrate nuclear's important contribution to clean energy. Not everyone is anti-nuclear.

Environmentalists

What's the current opinion among environmentalists on nuclear energy? Nicholas Stern's important *Stern Review* (2007) advised the UK government on economics and climate change; he subsequently stated on GHG emissions' policy: "The challenge is to produce an effective, efficient and

---

37. Adams, "Continuing Education on Advanced Reactors." The video of the hearing is informative.
38. *WNN*, "Summit Urges Action."
39. Hodson, *Donald Trump*.
40. Mecklin, "Introduction: Nuclear Power," 1.
41. Stover, "Kerry Emanuel."

equitable set of principles and policies to guide both national action and a global deal."[42] Here he recognized: "Nuclear power and CCS both have powerful critics within environmental communities who are arguing for strong action on climate change. My own view is that we will need all the technologies available. Each has different pros and cons."[43] I concur; we need all technologies.

"Green groups" using an anti-nuclear narrative often reject nuclear power. Friends of the Earth (FOE) Cymru in *Evidence to the Welsh Affairs Committee* (2006) recognized that the challenge of climate change requires reductions in dioxide emissions. But it is unequivocal: "Friends of the Earth Cymru opposes the construction of a new generation of nuclear reactors because a range of safer, greener and cleaner alternatives can deliver greenhouse gas reductions to meet climate change targets and maintain energy security."[44]

Moreover, FOE's 2015 *Nuclear Power* stated: "We do not support the building of new nuclear power plants in the UK."[45] Also, they oppose nuclear plant life extensions if "any significant safety concerns" arise or they "crowd out renewables." Nonetheless, FOE don't oppose research into nuclear plant using thorium fuel, seen as potentially safer. *Nuclear Power* utilizes commissioned research with the Tyndall Centre for Climate Change Research (Manchester University), which recommends regular reviews in case technology doesn't develop as anticipated.[46] Their recommendation is sensible.

In 2007, Greenpeace International issued *Climate Change - Nuclear Not the Answer*. It starkly stated: "For the planet and its people we must all make the right choice we must choose efficient and safe renewable energy sources over dirty and dangerous nuclear power."[47] Their contrast is clear, but we should ask: Can't we have both renewables and nuclear? Greenpeace's current campaign rejects nuclear "because it is an unacceptable risk to the environment and to humanity."[48] When Greenpeace's Doug Parr gave evidence to the Welsh Affairs Committee on "The future of nuclear power

---

42. Stern, *Blueprint for a Safer Planet* (2009), 99. Lord Stern was Chief Economist and Senior Vice-President of the World Bank (2000–2003). When he wrote *Blueprint*, he held the position of I.G. Patel Chair at the London School of Economics.

43. Ibid., 123.

44. FOE Cymru, "Evidence," 8.

45. FOE, *Nuclear Power*. For their detailed discussion see *Plan for Clean British Energy*.

46. Tyndall Centre, *Review of Research* (2013).

47. Greenpeace, *Climate Change*, 1.

48. Greenpeace, "End the Nuclear Age." They want to prevent nuclear expansion and close existing reactors.

in Wales" (March 2016) he reasserted: no new nuclear plant at Wylfa.[49] But surprisingly he included gas in his energy mix. Greenpeace pits nuclear and renewables as enemies, not allies—a future exists with renewables but not nuclear.[50] Many people consider such polarization unnecessary.

In contrast to a negative nuclear narrative there are pro-nuclear environmentalists, including people who have changed their allegiance. Anglican Bishop Hugh Montefiore, an environmentalist, converted to a pro-nuclear position. Michael de-la-Noy's 2005 *Guardian* obituary observes he retired in 1987 but continued with environmental issues until 2004, "when he was forced to resign from the board of FOE (of which he had been chairman from 1992 to 1998) after promoting the use of nuclear power in the fight against global warming."[51]

In the *Independent* (October 2004) Montefiore outlined his position: "I've been a Friends of the Earth trustee for 20 years, but I am told it is incompatible with being pro nuclear energy."[52] He argued that, as a theologian and a long-time committed environmentalist, the gravity of global warming meant he saw a solution in more nuclear energy.

The environmentalist Stewart Brand considers that debate is necessary to determine what is right within changing circumstances, and noted: "In the face of climate change, everybody is an environmentalist."[53] Furthermore, Brand frankly admits his change of mind:

> *Unfortunately* for the atmosphere, environmentalists helped stop carbon-free nuclear power cold in the 1970s and 1980s in the United States and Europe. (Except for France, which *fortunately* responded to the '73 oil crisis by building a power grid that was quickly 80 percent nuclear.) Greens caused gigatons of carbon dioxide to enter the atmosphere from the coal and gas burning that went ahead instead of nuclear. I was part of that too, and I apologize.[54]

---

49. HCWAC, "Future of Nuclear Power in Wales." See also FOE Cymru, *Submission to the Welsh Affairs Committee* (2016).

50. Greenpeace, *Renewable Energy vs Nuclear Power*; and Teske et al., *100% Renewable Energy for All*.

51. de-la-Noy, "Rt Rev Hugh Montefiore."

52. Montefiore, "We Need Nuclear Power."

53. Brand, *Whole Earth Discipline*, 1.

54. Ibid., 17–18. Italics original.

Canadian Patrick Moore, a cofounder (1971) and president (1977) of Greenpeace, describes himself as "the sensible environmentalist."[55] Moore left Greenpeace in 1986. He explained: "Since I left, Greenpeace has adopted many positions, including hanging on to the mistake of being against nuclear energy, that I do not agree with from an environmental perspective."[56]

Another convert to the benefits of nuclear is American Gwyneth Cravens. "In 1980 she was among the activists who shut down the $6 billion Shoreham Nuclear Power Plant in Long Island before it ever opened, which helped frighten the American nuclear industry to a standstill."[57] However, in the 1990s she met Dr. "Rip" Anderson, who shared his view from within the nuclear industry. Their travels through the US nuclear industry resulted in her engaging 2007 book *Power to Save the World: The Truth about Nuclear Energy*—a remarkable testimony to her journey.

Finally, British environmentalist and pro-nuclear advocate James Lovelock, from his first book (1979) showed his early support for nuclear energy. He answered fellow environmentalists' objections.[58] It must be noted that Lovelock is not a recent convert to nuclear energy, although some environmentalists seem to think it. Lovelock asserted this is untrue—his first book shows his early support.[59]

Mark Lynas, a British environmental writer and campaigner, and formerly anti-nuclear activist, wrote about his conversion in the *New Statesman* (May 30, 2005).[60] In a 2013 BBC Politics video Lynas visits the operational nuclear power plant at Hinkley Point B in Somerset (a plant where I worked in the 1970s) and the BBC commented: "Environmentalist Mark Lynas said he 'grew up hating nuclear power' but later realized that 'continuing to oppose nuclear was a mistake.'"[61] His book *Nuclear 2.0* is a compelling case for nuclear energy. Here he explains his journey from anti-nuclear environmentalist to realizing, during an energy conference at Oxford University (2005), he had overlooked the value of nuclear as a

---

55. http://www.ecosense.me.

56. Moore, "Patrick Moore." Part 2 of the interview with Joseph Cotto of the *Washington Times* in 2012. See also Moore, "Going Nuclear."

57. Brand, *Whole Earth Discipline*, 80.

58. See, for example, Lovelock, *Revenge of Gaia*, 87–105.

59. Ibid., 91.

60. Lynas, "Nuclear Power."

61. Lynas, "Nuclear Power Support."

low-carbon technology. He "was stunned," noting: "As an environmentalist I had come of age within a movement that regarded anything 'nuclear' as irredeemably dangerous and evil, yet its potential to tackle climate change was undeniable."[62] He began researching and writing on nuclear's "real risks and benefits."[63]

In 2009 the *Independent* newspaper reported on four leading environmentalists, former active opponents of nuclear energy, who changed their stance in a major u-turn to pro-nuclear because of the dangers of climate change from fossil fuels. These were: Stephen Tindale (UK Executive Director of Greenpeace, 2000–2005), Lord Chris Smith of Finsbury (Chairman of the Environment Agency, 2008–2014), Mark Lynas, and Chris Goodall. Tindale commented that his change "was kind of like a religious conversion. Being anti-nuclear was an essential part of being an environmentalist for a long time." For him discussions with other environmentalists now confirm "that nuclear power is not ideal but it's better than climate change."[64] Lynas sees environmental campaigners' anti-nuclear stance "as an enormous mistake" for the climate, and, "To give an example, the environmentalists stopped a nuclear plant in Austria from being switched on, a colossal waste of money, and instead [Austria] built two coal plants."[65] Stephen Tindale is cofounder of *Climate Answers*. The film *Pandora's Promise* features leading environmentalists supporting nuclear energy.[66] Finally, *An Ecomodernist Manifesto* (April 2015) endorses nuclear: "Nuclear fission today represents the only present-day zero-carbon technology with the demonstrated ability to meet most, if not all, of the energy demands of a modern economy."[67]

Surprising, after the Fukushima accident *Guardian* newspaper columnist and environmentalist George Monbiot declared himself "no longer

62. Lynas, *Nuclear 2.0*, 8.

63. Ibid. In *Nuclear 2.0* he helpfully shares his views, supported by figures, to inform and challenge others. He supports using both renewables and nuclear energy to tackle climate change.

64. Cited in Brand, *Whole Earth Discipline*, 90. Also, cited in Connor, "Nuclear Power." On his recent views see http://climateanswers.info/2016/12/director-stephen-tindales-evidence-to-lords-committee/.

65. Connor, "Nuclear Power?"Brackets original.

66. Director Robert Stone is a convert to nuclear energy. See his 2016 blog post in *Scientific American*: Stone, "Education of an Environmentalist."

67. Asafu-Adjaye et al., *Ecomodernist Manifesto*, 23. The authors suggest that challenges make it unlikely that current nuclear technologies will give significant climate change mitigation. New generation technologies (safer and cheaper) are likely to be needed for nuclear energy to reach its full potential in mitigating climate change.

nuclear-neutral," but a supporter of nuclear energy, explaining his change of heart and position.[68] Since then he has become an active pro-nuclear supporter.

My view, as a scientist and Christian, is that the UK needs energy efficiency savings, reduced consumption, renewables, and nuclear in our future energy mix to fulfill the criteria of (a) affordability, (b) energy security, and (c) a low-carbon economy.[69] Because climate change is a global issue countries should continue to use nuclear energy and build new nuclear plants for clean low-carbon energy.

UK Nuclear Policy

For many years in the UK there has been an ongoing debate about nuclear issues. Indeed, the Royal Society and the Royal Academy of Engineering established a joint working group on the future of nuclear energy. Their 1999 report took a pro-nuclear position:

> It is vital to keep the nuclear option open. We cannot be confident that the combination of efficiency, conservation and renewables will be enough to meet the needs of environmental protection while providing a secure supply of electricity at an acceptable cost. It is essential to win back public confidence in this option.[70]

The working group endorsed the House of Commons Trade and Industry Committee's 1998 recommendations of presuming that new nuclear would be required.[71]

Initially, the Labour government's 2003 white paper emphasized energy efficiency and renewables; new nuclear was neglected.[72] But later a Greenpeace briefing captured a changed situation.[73] Greenpeace saw the nuclear lobby and industry as responsible for the government's change of heart, and warned that new nuclear was the wrong answer for climate

---

68. Monbiot, "Going Critical."

69. The criteria are often called the energy "trilemma" of meeting cost, security, and decarbonization. See the definition: RAE, *Critical Time for UK Energy Policy*, 9–10.

70. RS and RAE, *Nuclear Energy*, 3. Moreover, the same page stated: "It is not appropriate to dismiss an energy source on the grounds that it could supply 'only' a few percent of need."

71. Ibid., 3. The summary records the main conclusions, endorsed by both councils (4).

72. DTI, *Our Energy Future*.

73. Greenpeace, *2005 Energy Review*.

change and energy security—renewables were needed. But again I ask: Can't the UK's energy mix have both? New Labour's 13 years in office saw no new nuclear plant constructed.[74]

With the later coalition government (Conservatives and Liberal Democrats) nuclear strategy was developed. The Conservative government (elected 2015) continued this commitment. It is beyond the scope of this book to describe this detailed work, but I offer an overview.[75] First, in 2011 the Department of Energy and Climate Change (DECC) published the *Overarching National Policy Statement for Energy (EN-1)*, the first of technical documents for the energy sector, including nuclear generation.

Then the UK's 2011 *National Policy Statement for Nuclear Power Generation (EN-6)* stated:

> The Government believes that energy companies should have the option of investing in new nuclear power stations. Any new nuclear power stations consented under the Planning Act 2008 will play a vitally important role in providing reliable electricity supplies and a secure and diverse energy mix as the UK makes the transition to a low carbon economy.[76]

Importantly, these new reactors will not be state owned but privately owned and operated. This means the UK government promotes an energy policy but doesn't actually build plants, although it was urged to move quickly to encourage long-term investment etc.[77]

Government vision and reality don't always coincide. For example, the House of Lords Science and Technology Committee's 2011 publication *Nuclear Research and Development Capabilities* criticized the government. Their task was not to examine the pros and cons of nuclear energy, but to consider whether the government was doing enough considering its declared intention (of providing future electricity which is secure, affordable, and low-carbon). The Committee commented:

---

74. Although a 2008 white paper came out supporting new nuclear builds with Gordon Brown (the Labour Prime Minister after Tony Blair) supporting nuclear.

75. For more on politics, see Giddens, *Politics of Climate Change*; and Dobson, *Environmental Politics*.

76. DECC, *National Policy Statement*, 1, para. 1.1.1.

77. IMechE, "Nuclear Build," 5, gives key recommendations. On the UK's historical context and the move away from stated-owned nuclear power plants, see Nuttall and Earp, "Nuclear Energy in the UK." Also, see Grimston et al., "Siting of UK Nuclear Reactors," R15.

> Nuclear energy currently supplies 16% of the UK's electricity (10–12 gigawatts (GW) of capacity). Scenarios for future electricity generation suggest that between now and 2050 nuclear power could supply between 15% and 49% (12 and 38 GW) of the total. To meet the UK's legally binding target of reducing greenhouse gas emissions to 80% below 1990 levels by 2050 it is likely that between 20 and 38 GW of nuclear power will be needed.[78]

Was the government doing enough? The Committee "concluded that they are not."[79] The government's vision was failing in areas of leadership, strategic thinking, research and development, and expertise (past achievements may dwindle as experts retire). The government was held to be complacent. The Select Committee produced conclusions and recommendations for government action.

A 2013 government report asserts nuclear will continue as:

> a key part of our low-carbon energy mix alongside renewable generation and Carbon Capture and Storage. All of these technologies are important in tackling climate change and diversifying our supply, contributing to the UK's energy security and growth.[80]

It thus reasserts the role of nuclear power, addressing the ambitious challenges ahead including partnership between government and industry. I agree with this approach.

Amber Rudd's November 2015 speech on UK energy policy reaffirmed the need for national energy security which includes energy efficiency, gas, renewables, and nuclear. Coal is to be phased out.[81] Yet there was no plain sailing for the government. For example, Stephen Tindale's webpage praised Rudd's speech but less than a month later in December 2015 commented on climate and energy policy: "small steps forward, large steps backwards," after the Chancellor made cuts.[82] Since then the government has instituted a competition for the development of small modular reactors (SMRs) and

---

78. HLSTC, *Nuclear Research*, 6. Footnotes omitted but 16 percent refers to 2010. The range of 15–49 percent depends on assumptions and is just an indication.

79. Ibid., 3.

80. BIS, *Nuclear Industrial Strategy*, 3. Diversifying supply ensures the UK has multiple sources of generation to ensure ongoing energy security. For more on diversification see the video Nuttall, "Britain, Nuclear Energy."

81. Rudd, "Amber Rudd's Speech." Rudd is former Secretary of State for Energy and Climate Change.

82. Tindale, "Well Done Amber Rudd" and "UK Climate and Energy Policy."

a Department for Business, Innovation & Skills (BIS) press release (May 9, 2016) confirms a UK grant of £80 million will boost training, including £15 million for nuclear training colleges.[83] The House of Commons Welsh Affairs Committee (HCWAC) recommends that Trawsfynydd is designated as a site for a SMR.[84]

In March 2017 the House of Lords Science and Technology Committee's report *Nuclear Research and Development Capabilities* (2011), mentioned above, was revisited on some of its conclusions and recommendations (and government responses) under the title *Priorities for Nuclear Research and Technologies*. The Science and Technology Committee's January 2017 Call for Evidence announced it would look at SMR design for the UK, the role of the National Nuclear Laboratory, and the Nuclear Research and Advisory Board (NIRAB). In May 2017 the Science and Technology Committee published *Nuclear Research and Technology: Breaking the Cycle of Indecision*. This reiterated some of its 2011 recommendations, made further recommendations, and remarked that "The undoubted potential of civil nuclear has been blighted by the indecision of successive Governments" (3). Urgent government action was required. Its "Summary of Conclusions and Recommendations" is worth reading. This committee shows the importance given to the civil nuclear industry and the need to develop an appropriate long-term strategy.[85]

Moreover, the House of Commons Business, Energy and Industrial Strategy Committee addressed *Leaving the EU: Negotiation Priorities for Energy and Climate Change Policy Inquiry* and its report appeared in May 2017.[86] On February 28, 2017, nuclear industry members appeared as witnesses before the committee when it investigated "Leaving the EU: energy and climate negotiating priorities." Among the questions addressed was the current role of the European Atomic Energy Community (Euratom) and how the UK would function once it withdrew from Euratom in the process of Brexit. This exit from Euratom has been analyzed by the Institution of Mechanical Engineers, which stated: "This paper begins to explore the implications of

83. BIS, "Government Confirms £80 Million." For more on SMRs, see my ch. 7.
84. HCWAC, *Future of Nuclear Power*, 34–39, 47.
85. HLSTC, *Priorities for Nuclear Research and Technologies*. A video recording of evidence presented by nuclear industry witnesses can be viewed at http://parliamentlive.tv/Event/Index/17ed7c54-3ed9-42fb-9fcc-e56a57af3555. HLSTC, *Nuclear Research and Technology*.
86. HCBEISC, *Leaving the EU* and *Oral Evidence: Leaving the EU*. The report *Leaving the EU: Negotiation Priorities for Energy and Climate Change Policy* (HC 909) includes many recommendations.

Euratom departure, the opportunities and threats created, and recommendations towards a strategy safeguarding UK nuclear interests."[87] Since no country has previously left the European Union, the UK will enter uncharted waters, but it is aware of what must be achieved one way or another.

But Hinkley Point C new nuclear build remains controversial, even among pro-nuclear advocates, on the grounds of electricity costs, finance, and delays. So, what can we say? It is a long process building new nuclear plants, but partnerships continue working. The government seems determined to deliver suitable circumstances for new nuclear build. This is important since a 2016 survey by the Institution of Mechanical Engineers sees an UK electricity gap emerging and we don't want power cuts, do we?[88]

In September 2016 the UK government under Prime Minister Theresa May finally approved the construction of Hinkley Point C nuclear plant—a major decision. Furthermore, since the country voted in June 2016 to leave the European Union (Brexit) the impact on the UK's future energy situation came under discussion.[89]

In February 2017, the UK government issued its white paper: *The United Kingdom's Exit from and New Partnership with the European Union*.[90] This confirmed that Brexit meant leaving the EU and the European Atomic Energy Community (Euratom). Much discussion has centered on concerns over leaving Euratom, its many impacts on the nuclear industry (e.g., safeguards, nuclear trade, resources, and fusion research), and the negotiations for alternative arrangements.[91] In the white paper, Prime Minister Theresa May confirmed: " . . . the nuclear industry remains of key strategic importance to the UK and leaving Euratom does not affect our clear aim of seeking to maintain close and effective arrangements for civil nuclear cooperation, safeguards, safety and trade with Europe and our international partners."[92] Subsequently, the Brexit Bill passed through both Houses of

---

87. IMechE, *Leaving the EU*, 1. Also see NIA, *Exiting Euratom* (May 2017).

88. IMechE, *Engineering the UK Electricity Gap*.

89. WNN, "UK Parliamentary Hearings," August, 2016.

90. HMG, *United Kingdom's Exit*. A white paper is a report that provides information.

91. Leech and Cowen, "Brexit White Paper Confuses." For valuable practical advice see the Nuclear Institute's April 2017 response: NI, *Brexit and the Euratom Treaty Issue*.

92. HMG, *United Kingdom's Exit*, 44, para. 8.31. The May 2017 report HCBEIS *Leaving the EU* (HC 909), 53–54, warned "*We note that the necessity of leaving Euratom is subject to legal uncertainty, but that uncertainty puts the continuing operation of the nuclear industry in the UK at risk. The Government therefore has a responsibility to resolve this matter as urgently as possible. (Paragraph 85)*". Italics original.

Parliament and on March 29, 2017, Theresa May triggered Article 50, the formal process for the UK to leave the EU. Negotiations between the UK and the EU occur over a period of two years. Although the UK moves into unchartered waters, the government's commitment to nuclear energy continues, but it, and successor governments, must urgently act on the recent reports and recommendations of the Science and Technology Committee and the House of Commons BEIS Committee. *World Nuclear News* succinctly reported "UK Nuclear's Future in Government Hands, Say Reports." The Nuclear Innovation and Research Advisory Committee (NIRAB) also produced a relevant final report in February 2017.[93]

Political Parties

Let us briefly see where other parties stand. The Labour Party's 2015 manifesto comments: "We will create an Energy Security Board to plan and deliver the energy mix we need, including renewables, nuclear, green gas, carbon capture and storage, and clean coal."[94] The Liberal Democrats' 2015 manifesto notes that they "Accept that new nuclear power stations can play a role in low-carbon electricity supply provided concerns about safety, disposal of waste and cost are adequately addressed and without public subsidy for new build."[95]

The Green Party recognizes the dangers of climate change and responds with policy which supports a transformation of the UK's energy system through energy efficiency and primarily using renewables.[96] Included among practices that it opposes is all nuclear power generation, so construction of new nuclear power plants will be cancelled.[97] Its 2015 manifesto states that it opposes nuclear and will "Phase out nuclear power within ten years."[98]

---

93. HCSTC, *Nuclear Research and Technology*, and HCBEIS, *Leaving the EU* (HC 909). NIRAB, *NIRAB Final Report: 2014–16*. NIRAB was formed in January 2014 for a period of three years to advise Ministers on civil nuclear research that was publicly funded. To fulfill its role it worked with universities, national laboratories and industry. It was an independent advisory board.

94. Labour Party, *Britain Can Be Better* (2015), 20.

95. Liberal Democrats, *Stronger Economy, Fairer Society*, 28.

96. Green Party, "Getting Britain Working," paras. EN007 and EN001.

97. Ibid., EN261. See also EN262: money allocated to new nuclear build including research and development will be reallocated. However, funding for decommissioning nuclear plants and radioactive waste is retained.

98. Green Party, *For the Common Good*, 23. On opposing nuclear energy see p. 24.

The Scottish government had responded, in 2007, to the UK government's consultation on the future of nuclear energy.[99] While recognizing climate change challenges and energy security needs, its position stated these could be achieved "without the development of new nuclear power stations." The government argued that developing technologies (e.g., renewables, clean fossil fuel, and carbon capture and storage) would be better for energy security than nuclear. Nevertheless, more recently, it stated: "Scottish Ministers have made it clear that they are supportive of possible life extension of existing nuclear power stations in the short term to help security of supply."[100] But it is against new nuclear power plants. The Scottish National Party's 2015 manifesto was surprisingly silent on its two operating nuclear plants.[101] However, EDF Energy is supporting life extensions for these plants.

Unlike Scotland, which still has two nuclear power stations operating, Wales' two nuclear power plants (at Wylfa and Trawsfynydd) are permanently closed down. In 2016 the House of Commons Welsh Affairs Committee (HCWAC) launched an enquiry into the future of nuclear power in Wales and invited evidence (see more below).[102] The UK government proposes to build new reactors at both sites.[103] Wales currently remains open to nuclear energy but this is conditional. For example, the Committee stated:

> The UK Government is in favour of new nuclear build, but not at any price. Energy policy should balance cost against energy security and environmental concerns. We recommend that the Government negotiate a strike price for Wylfa Newydd below that agreed for Hinkley Point C and seek a price that would be competitive with renewable sources, such as on-shore wind.[104]

So, nuclear has qualified political support. The House of Commons Science and Technology Committee's (HCSTC) report *Devil's Bargain? Energy Risks and the Public* recognizes:

---

99. Scottish Government, "Scottish Government's Response."

100. Scottish Government, "Business, Industry & Energy," answering the question: "What are the main reasons for Scottish opposition to nuclear power?".

101. SNP, *Stronger for Scotland*, 5–6. The SNP, *Re-Elect Manifesto 2016* supported the current plants but banned new nuclear plants because of costs.

102. HCWAC, *Future of Nuclear Power*, 7. This report, published in July 2016, is an excellent source of information, sharing and weighing the evidence taken at the inquiry.

103. At Wylfa the proposal is to have two UK Advanced Boiling Water Reactors (ABWRs) and at Trawsfynydd small modular reactors (SMRs).

104. HCWAC, *Future of Nuclear Power*, 41, para. 2. The committee stated that if the price was too high the government should cancel the project.

The Government's position as an advocate for nuclear power makes it difficult for the public to trust it as an impartial source of information. However, regulatory bodies that are independent of Government and technically competent are in a unique position to engender public trust and influence risk perceptions.[105]

At this stage let us turn to the role of the regulators.

The Regulators

Regulation and regulators perform an important role in the nuclear industry. During my master's degree studies in health physics (radiation protection) I learnt about regulations/compliance; then when I joined a nuclear power station I met regulators/inspectors. We ran staff emergency training exercises with nuclear inspectors assessing our performance. They asked questions; we provided answers. The regulator's role was essential then as it is now.

In 2006 the UK government asked regulators to consider "pre-authorisation" assessments for new nuclear plants.[106] They produced a four-step Generic Design Assessment (GDA) process. All new reactor designs are submitted by the vendor company and are expected to complete the GDA (note that this is a voluntary process based on policy and not law), so it's worth outlining the steps specified by the Office for Nuclear Regulation (ONR):

- Step 1 is the preparatory phase, setting up formal agreements and discussions.

- Step 2 is an overview of the fundamental acceptability of the reactor design in the UK's regulatory system. It aims to identify "design or safety shortfalls" and key assertions to see if the reactor is acceptable for UK construction.

- Step 3 continues the step 2 work of examining assessments of safety arguments and claims. It also considers security.

- Step 4 analyzes evidence presented on safety case claims and arguments. At the end of step 4 the regulator determines whether enough evidence has been submitted for acceptance of the design.[107]

---

105. HCSTC, *Devil's Bargain?*, 3.
106. Sciencewise, *New Nuclear Power Stations*, 1.
107. ONR, *Summary Report of the Step 3*, 4–5.

This process means key features (safety, security, and environmental) are addressed at the design stage and before construction. Any regulatory "queries," "observations," or "issues" are fed back to the provider for further information or work during the steps.[108]

Initially, four companies (requesting parties) indicated their interest in GDA when the process started in July 2007. Subsequently, only one company continued at that stage. The first new plant to complete GDA was the UK EPR and in December 2012 the ONR granted it a Design Acceptance Confirmation (DAC) while the Environment Agency (EA) provided a Statement of Design Acceptability (SoDA).[109] By March 2013 all technical assessment reports were published, bringing the process to completion.

At the time of writing, Westinghouse's AP1000 and Hitachi-GE's UK Advanced Boiling Water Reactor (UK ABWR) were going through GDA. The intention is to build the ABWR at Wylfa Newydd in Wales and then Oldbury (near Bristol) in England. The ABWR has completed step 3 and ONR have published their report.[110] It demonstrates the significant amount of work achieved by the regulators with Hitachi-GE and Horizon Nuclear Power. Nevertheless, much work remains as they start step 4, with a planned completion in December 2017—meanwhile in April 2017 Horizon Nuclear Power announced that it had applied to the ONR for a Wylfa Newydd site licence.[111] The AP1000 had an expected completion date for the GDA process of March 2017, which was achieved when the ONR issued its Design Acceptance Confirmation (DAC) while the Environment Agency (EA)

---

108. For an explanation of each term, see ONR, "Frequently Asked Questions," Q20.

109. The UK EPR is the European Pressurised Reactor which EDF Energy is building at Hinkley Point C. Two reactors will generate 3,260 MWe for a 60-year service life. It uses low-enrichment fuel as uranium dioxide pellets enclosed in a fuel assembly. This is cooled by pressurized light water that also acts as a moderator for neutrons. The hot water heats a secondary circuit (steam generators) which produces steam to drive the turbine.

110. ONR, *Summary Report of the Step 3*. ABWRs uses enriched uranium fuel, cooled by pressurized light water which is also the moderator for the neutrons. The water is boiled in the reactor and this steam is taken directly to the turbine to produce electricity. Unlike pressurized water reactors (PWRs), ABWRs do not have a secondary water/steam system linking the reactor and turbine. The planned plant life is 60 years. Each reactor generates 1,350 MWe. For more information, see Hitachi-GE's website: http://www.hitachi-hgne-uk-abwr.co.uk/index.html.

111. See HCWAC, *Future of Nuclear Power*, 19–20, on the GDA for Wylfa Newydd (also called Wylfa B). On the nuclear site licence application, see: *WNN*, "Horizon Applies for Wylfa Newydd Site Licence."

provided a Statement of Design Acceptability (SoDA).[112] The company NuGen plans to build the AP1000 at Moorside in Cumbria (subject, of course, to all requirements being met). Then, in January 2017 the British government asked the regulators (the ONR and the Environment Agency) to start a GDA for the Chinese design reactor UK HPR1000 (Hualong One). The proposal is to build this new reactor at Bradwell in Essex.[113]

The implementation of GDA is a valuable structured approach paid for by the company. It is open and transparent, inviting public input.[114] This is important. Companies display information on their websites for public access.[115] Regulators work independently of the government—an essential point. Their mission statement crucially includes public involvement:

> "... holding the industry to account on behalf of the public", and we place great importance on being open and transparent about our work and the regulatory decisions that we make. We believe that this will help to improve and maintain public trust in the work that we do. ONR publishes all of its reports, statements and guidance on the joint regulators' GDA website, which includes an electronic news bulletin specifically for those interested in new nuclear reactors.[116]

In the UK, "passing" the GDA is not sufficient. Before construction and operation new nuclear plants require site-specific key permissions, including:

- planning permission from the government;
- a nuclear site licence issued by the Office for Nuclear Regulation (ONR);
- environmental discharge permits from either the Environment Agency (EA) in England or Natural Resources Wales (NRW).[117]

---

112. ONR et al., *Assessing New Nuclear Reactor Designs* (a quarterly report for February–April 2016) provides a detailed progress update on the GDA. See ONR, "Advanced Passive 1000—AP1000®," and *WNN*, "AP1000 Design Completes."

113. For general information, see *WNN*, "UK to Start." On the design philosophy and technical aspects, see Xing et al., "HPR1000." Ji Xing and coauthors work for the China Nuclear Power Engineering Company in Beijing.

114. See ONR, "Frequently Asked Questions." Openness and transparency were covered in an EA inspector's presentation at a Nuclear Institute meeting I attended on April 18, 2016, in Bristol: "Regulating Nuclear Sites and New Nuclear Build—the Environment Agency's Role."

115. The ONR publishes quarterly updates (http://www.onr.org.uk/new-reactors/quarterly-updates.htm), which can be emailed directly to interested people on request.

116. ONR, *Summary Report of the Step 3*, 13.

117. ONR, "Frequently Asked Questions," Q15. For further information on

ONR's role does not stop here but continues into construction of the nuclear plant, operation, maintenance, and decommissioning. It carries out regular site inspections to ensure the nuclear site licence conditions are kept, suggests improvements in safety, and assesses site emergency procedures and demonstrations. Moreover, it has enforcement powers from offering advice to prosecutions in court.[118] Such information is published on its website, so no accidents are "hushed up." Also, the International Atomic Energy Agency (IAEA) conducts periodic in-depth inspections of the UK's nuclear power plants.[119] So the UK's new nuclear plants, like current plants, are well regulated.

Nevertheless, media reports have expressed concern that the number of ONR inspectors is insufficient for regulation.[120] In response, ONR provided the reassuring comment: "Recent media articles have suggested that ONR is not sufficiently resourced to meet current industry demands for effective regulation. That is not the case."[121]

Universities/Colleges and Employers

Many UK universities support the nuclear industry, e.g., through undergraduate and postgraduate courses. Research centers and groups also provide support (e.g., Dalton Nuclear Institute, University of Manchester).[122]

---

requirements for a new nuclear plant, particularly on public consultations, see http://www.horizonnuclearpower.com/wylfa-newydd?area=wylfa and http://consultation.horizonnuclearpower.com for various resources/reports.

118. See HCWAC, *Future of Nuclear Power*, 19, for ONR's role once a reactor is approved in the UK, and ONR, *Enforcement Policy Statement*. For examples of enforcement, see ONR, *Regulation Matters* (July 2016), 6.

119. An inspection occurred at Sizewell B in October 2015. See IAEA, *Report of the Operational Safety Review Team*.

120. For example, in *The Times*: Pagnamenta, "Ageing Nuclear Plants," 35. Also see Mellen and Hollow, *No Oil*, 48.

121. ONR, "ONR Staff Recruitment." Early in 2016 the ONR's website displayed staff recruitment adverts for inspectors. Moreover, in evidence to the Welsh Affairs Committee, ONR confirmed it was increasing the "inspectorate team" from 350 (now) to a projected 550 (by 2020) to cover further GDA applications. HCWAC, *Future of Nuclear Power* (July 2016), 35, para. 130.

122. Dalton Nuclear Institute: http://www.dalton.manchester.ac.uk. There is also an annual UK nuclear academics discussion meeting. In early September 2016, this met for two days at the University of Bristol. See Roberts and Grimes, "2016 Nuclear Academics." The valuable presentations from this meeting are available to download at Nuclear Universities (http://www.nuclearuniversities.ac.uk).

Partnerships also exist. For example, in 2012 the University of Birmingham published the comprehensive report *The Future of Nuclear Energy in the UK*, which examined the outstanding challenges, present and future, to see if nuclear energy had a future role. The Birmingham Policy Commission's authors have had extensive expertise in the nuclear industry, drew upon a wide range of contributors and also engaged with people holding anti-nuclear views.[123] They concluded: "unless the Government shows a decisive lead and creates the right conditions for investment in the UK, the country risks losing out on its huge potential for developing a new nuclear industry."[124]

In March 2015 Business Minister Matthew Hancock announced that four education providers were partners in a virtual National College for Nuclear (NCfN). It specializes in training and development for the industry and addresses skills shortages. Hancock commented:

> It's expected the nuclear industry will need 30,000 new employees over the next decade—and the Nuclear College will equip young people with the skills they need. Creating jobs and opportunities for local people is front and centre of our long term economic plan to secure a brighter future for Britain.[125]

Subsequently, the government announced £15 million government funding for NCfN.[126]

Specifically, the University of Bristol operates as a South West Nuclear Hub with Oxford University, arising from their earlier Joint Nuclear Research Centre partnership.[127] This academic component operates an interface linking the nuclear industry and government agencies. A new Master of Science (MSc) in Nuclear Science and Engineering degree program had its first intake in 2015, and in May 2016 Bristol University ran a two-day "Seminar on the ABWR and Underpinning Nuclear Energy R&D," which I attended.[128] During this seminar I had the opportunity to take a tour and

---

123. Birmingham Policy Commission, *Future of Nuclear Energy*, 9.

124. Ibid., 2.

125. Bridgwater and Taunton College, "National College for Nuclear"; and University of Bristol, "Bristol Named." By 2025 it is estimated that 70 percent of the current UK's nuclear personnel will have retired. With decommissioning, current operating reactors, and new nuclear builds a skills gap is identified. See, e.g., Cole et al., "Strategies," N25–N26.

126. University of Cumbria, "BIS Announces £15 Million" (May 2016).

127. Scott, "Meeting the Skills Challenge."

128. The seminar (May 10–11, 2016) was part of series run by Hitachi-GE Nuclear Energy on the ABWR which Horizon Nuclear Power plans to build in the UK.

see valuable research work on nuclear safety that the university is conducting in partnership with industry.[129] In September 2016 the university hosted the UK Nuclear Academics Meeting.[130]

Moreover, there is a UK Nuclear University Network (http://uk-nuclear.net) cosponsored by the Nuclear Institute and the National Skills Academy. An MSc in Nuclear Science and Technology course is delivered by seven universities plus the Defence Academy of the UK. Furthermore, Sir Andrew Witty's 2013 review of UK universities and growth identifies, besides education and research, an enhanced third mission: working with partners to "make facilitating economic growth a core strategic goal."[131] The nuclear industry is identified in the top 20 organizations (by publications), and these are overwhelmingly universities.[132]

The Open University (OU) has provided courses and books on nuclear power over many years.[133] In 2009 the National Skills Academy for Nuclear (NSAN) announced its formal association with the OU. The OU currently provides a range of courses on nuclear energy, including the free introductory online course "Unclear about Nuclear?" This is aimed at young people. The course leads to further studies.[134] Moreover, the OU offers research degrees in nuclear energy.[135] OU Professor of Energy William Nuttall explained the important role of nuclear energy in his 2015 lecture "Britain, Nuclear Energy and the Future."[136]

Besides universities, over 20 employers worked to produce two new approved apprenticeship standards (Nuclear Operative and Nuclear

---

129. On the university's nuclear research, see http://www.southwestnuclearhub.ac.uk/research/what-we-can-offer/.

130. It has UK senior academics, in nuclear energy topics, interacting nationally and internationally with institutions.

131. Witty, *Encouraging a British Invention Revolution*, 6. Economic growth refers to using universities' expertise to enhance regional economic growth, etc., and working with local enterprise partnerships.

132. Ibid., 103.

133. For example, Scott and Johnson, *Science Matters: Nuclear Power* (first published in 1993 with a 2nd edition in 1997), was part of the OU course Science Matters. Then in 2003 Boyle et al., eds., *Energy Systems and Sustainability*, formed part of the OU undergraduate course T206, Energy for a Sustainable Future.

134. OU, "Unclear About Nuclear?"

135. OU, "Nuclear Energy." Research is provided by a research council of three universities (Imperial College London, Cambridge University, and The Open University) with industry funding.

136. Nuttall, "Britain, Nuclear Energy." Lecture at the Royal Institution.

Scientist/Nuclear Engineer). The Nuclear Scientist/Nuclear Engineer route results in an accredited honors degree. These standards were supported by Sellafield Sites Ltd. and NSAN.[137] It's a valuable support system for developing skilled staff.

UK universities and colleges are centers for nuclear courses, research, development, and partnerships.

Trade Unions

Where does the UK Trade Union Congress (TUC) stand on nuclear energy? Well, its 2014 briefing "Unions 4 Climate Action" begins: "We can avoid climate related poverty, disease, unemployment and death."[138] It embraces "a just transition towards sustainable development," climate justice, and a willingness to work with others, e.g., religious communities. Consistent with science, it aims to decarbonize for today's workers and children. The TUC has 5.9 million workers and its national campaign aims for "a virtually carbon-free UK power sector by 2030," employing all energy sources—renewables, fossil fuels (with CCS) and new nuclear. The TUC supports the Hinkley Point C new build, with Paul Nowak, TUC Assistant General Secretary, visiting Hinkley Point's new union learning center (providing apprenticeships, etc.).[139]

In July 2016 the TUC issued a more detailed report comparing the UK with Germany and Denmark on energy and the environment. It reasserted its support for nuclear's pivotal role within the UK's energy mix but was critical of the government, progress over the construction of Hinkley Point, and called for a rethink to advance nuclear energy.[140] However, it recognized that compared to today's nuclear capacity (around 20 percent of the UK's electricity needs) a scenario of between 40 to 50 percent nuclear generated electricity by 2050 is possible.[141]

---

137. Nuclear Institute, "Trailblazing Apprenticeships for Nuclear," 8.

138. TUC, "Unions 4 Climate Action." Although the TUC made national commitments it did not specify the proportion of energy from renewables, fossil fuel, and new nuclear.

139. Unionlearn, "TUC Visits New Hinkley Point." He visited in January 2016.

140. TUC, *Powering Ahead*, 35–36. Its energy mix has: nuclear, renewables (at least 50 percent by 2030), developing CCS, and North Sea gas. Also see NIA, "Unions Back New Nuclear."

141. Ibid., 29. The scenario is one of 75 GW(e) of nuclear energy within a total installed capacity of 160 GW(e), drawing upon the government's carbon plan (2011).

Churches and Christians

We've considered politics and positions on nuclear power, so let us ask: How are Churches and Christians responding? There's no single answer. Besides opposition, there is acceptance (perhaps reluctant), silence, and suspicion. Perhaps complacency and confusion. Pope Francis noted, on the environment: "Obstructionist attitudes, even on the part of believers, can range from denial of the problem to indifference, nonchalant resignation or blind confidence in technical solutions."[142]

Churches and Christians recognize dangers in climate change and respond as part of Christian care for our world and mission.[143] Mission agencies (e.g., Tearfund and Christian Aid) report that climate change is happening now. Often, though, the benefits of nuclear energy are understated or objections surface to sideline them from serious discussion. Nevertheless, this isn't always the case, as I recognize. Christians (and others also) tend sometimes to dehumanize the nuclear industry. They fail to recognize the wide range of people working sincerely in the industry as people who love, serve, honor, and worship God. And God loves them.

Church of England

In 1982 the Church of England's Board of Social Responsibility commissioned a working party to produce a report which was entitled *The Church and the Bomb*. The working party agreed two terms of reference:

1. to study the implications for Christian discipleship of the acceptance by the major military powers of a role for thermo-nuclear weapons in their strategy;

2. to consider the bearing of this on the adequacy of past Christian teaching and ethical analysis regarding the conduct of war.[144]

The study addressed nuclear weapons, not nuclear energy. However, one subsection noted: "The fact that there are legitimate non-military reasons for enriching uranium or separating plutonium from spent nuclear

---

142. Pope Francis, *Laudato Si'*, 5. It is a comment on the environment and not specifically aimed at nuclear energy.

143. For example, see Archbishop John Sentamu's 2015 reflection, "Kiribati: Living in the Eye."

144. Baker et al., *Church and the Bomb*, vii–viii.

fuel is one reason why it is difficult to inhibit the horizontal proliferation of the capability to make weapons."[145] The authors understood the legitimacy of civil nuclear technology. This significant distinction is, unfortunately, not always acknowledged in current debates.

In 1984 the Anglican Communion agreed on its Five Marks of Mission. This important statement included the fifth activity: "To strive to safeguard the integrity of creation, and sustain and renew the life of the earth."[146] Caring for creation is a crucial part of mission. Then in 1990, while I was working as a tutor in the nuclear industry, a thoughtful collection of essays titled *The Power of God?* was published concerning the proposed Hinkley Point C reactor (which was not built, following the privatization of the Central Electricity Generating Board).[147] There was no overall negative response to nuclear. As we saw above, Bishop Hugh Montefiore became a convert to nuclear power and had to leave his position in Friends of the Earth.

In April 2013, during a debate in the House of Lords, the Bishop of Hereford spoke in favor of developing nuclear power.[148] Forward now to the 2015 General Synod debate "Combatting Climate Change" and mission. The Rt. Rev. Nicholas Holtam (Bishop of Birmingham) observed how the discussion had developed from earlier Synod discussions and since the 2005 resolution accepting the report *Sharing God's Planet* (by Claire Foster). Critically, he mentioned the earlier Five Marks of Mission, with the fifth: "To strive to safeguard the integrity of creation, and sustain and renew the life of the earth").[149] Therefore: "This is integral to evangelism and mission, not an 'it would be nice if.'"[150] I do agree.

The 2015 Synod debate motion began by recognizing the holiness of God's creation and the need to protect the Earth by ensuring average global temperatures do not exceed 2°C. The Synod sought to:

---

145. Ibid., 7. See the subsection "Civil and Military Uses of Nuclear Energy."

146. Anglican Communion, "Marks of Mission." http://www.anglicancommunion.org/identity/marks-of-mission.aspx.

147. Earney, *Power of God?* These thoughtful essays were published by the South West Region of the Industrial Mission Association and the Social Responsibility Group of the Diocese of Bath and Wells.

148. Caplin, "Bishop of Hereford Speaks."

149. Anglican Communion, "Marks of Mission." The 1991 classic book on mission by David J. Bosch *Transforming Mission* does not really address this mark of mission.

150. Church of England, *Report of Proceedings 2015*, 277.

a. urge all governments at the COP 21 meeting in Paris to agree long term pathways to a low-carbon future, supported by meaningful short to medium term national emissions pledges from all major carbon emitting nations;

b. endorse the World Bank's call for the ending of fossil fuel subsidies and the redirection of those resources into renewable energy options.[151]

During the debate (b) was addressed regarding nuclear energy. The Revd. Canon Dr. Christopher Sugden (Oxford) moved an amendment: "Some zero and low carbon sources such as nuclear and carbon storage are not renewables but are worthy of investment and do not involve regressive subsidies."[152] This produced a further aim to "encourage the redirection of resources into other lower carbon energy options."[153] The nuclear option was retained.

Also, in 2015 the Anglican Consultative Council and the Anglican Communion Environmental Network published the important declaration *The World Is Our Host: A Call to Urgent Action for Climate Justice*. It rightly recognized the human contribution to climate change and the role of fossil fuels but there was no statement for or against nuclear energy.[154] Furthermore, just before the papal encyclical and Paris climate talks in December 2015, the church issued the Lambeth Declaration on Climate Change (June 17, 2015), signed by the archbishops of Canterbury and York with other faith leaders. Its call on faith communities begins: "Recognise the urgency of the tasks involved in making the transition to a low carbon economy."[155] Many people see nuclear energy as part of this low-carbon economy.

Church of Scotland

Let us start with the Church of Scotland's report *Energy for a Changing Climate* (2007). What does it say? It focuses on electricity supply and energy saving, recognizing that generation and use of electricity comprises technical and economic assessments; but these alone are insufficient. Values

---

151. Ibid., 276. It then continued down to point (e).
152. Ibid., 289.
153. Ibid., 289, 290, and 299.
154. This declaration was signed by 17 Anglican bishops across the six continents and "sets a new agenda on climate change for the 85 million-strong Anglican Communion." Interfaith Power & Light, "17 Anglican Bishops."
155. Church of England, "Archbishop of Canterbury."

(with pros and cons) plus solutions have an ethical dimension (neglected by the government, it notes). In a section on theology there are two subsections: "Caring for God's Creation" and "General Ethical Principles." These principles are: (a) Responsible Stewardship of Creation; (b) Loving our Neighbour as Ourselves: Economic and Environmental Justice; and (c) Modifying our Lifestyles; Reducing our Consumption.[156] Under (b) they note: "being even handed in arguing the pros, cons, risks and benefits of rival technologies."[157] I think this is an important point and should provide overall guidance. In practice, they achieve this when assessing Scotland's options for future electricity generation. The report recognized that currently nuclear supplied "about half of Scotland's electricity," but the policy conclusions opted for energy savings, efficiency, and domestic renewables without new fossil fuel plants (even with CCS) or new nuclear.[158] They anticipate this mix will alleviate long-term electricity security concerns. Significantly, the problem is not just government ("them") but also "us."[159]

Its General Assembly of 2007 acknowledged "that we as a Christian family have failed in our stewardship of God's creation, and that we must now show leadership by considering the impact of even the most seemingly insignificant of our decisions."[160] This important admission of past failure and future possibilities is a model for critics of the nuclear industry's perceived failures—it's time to forgive and move forward.

Subsequently, the Society, Religion and Technology Project (SRTP) published "What Future for Nuclear Power?" (2010). Some members recognized diverse views within the church with some concerns over waste, accidents, and proliferation, while other members, balancing risks and impacts, suggested an ongoing role for nuclear energy. The SRTP argued: "The relatively low level of fossil-fired electricity in Scotland should not therefore be increased, but the existing nuclear component retained for the foreseeable

---

156. Church of Scotland, *Energy for a Changing Climate*, 5. Headings are explained using bullet points. Stewardship is an influential term often used by Christians, although its appropriateness is questioned. However, this is beyond the scope of my study. For a discussion, see Bauckham, *Bible and Ecology*, ch. 1, "Stewardship in Question."

157. Ibid. Under (b) this is point 6.

158. Ibid., 15, para. 5.6.1 and 19. However, in para. 6.3 the authors noted: "If pragmatically it is impossible to avoid some new building, we are not in a position to recommend which option we would regard as preferable."

159. Ibid., 20, para. 7.1.

160. Ibid., 25, point 16.

future."[161] Although it is not in favor of greatly increasing nuclear's share in the electricity market, it stated that old plants should be replaced by new build as necessary. The report encourages the development of nuclear fusion.

Roman Catholic Church

Pope Francis's Encyclical *Laudato Si'* advocates a new dialogue on environmental challenges:

> The worldwide ecological movement has already made considerable progress and led to the establishment of numerous organizations committed to raising awareness of these challenges. Regrettably, many efforts to seek concrete solutions to the environmental crisis have proved ineffective, not only because of powerful opposition but also because of a more general lack of interest.[162]

Although Pope Francis addresses energy efficiency plus renewable and non-renewable energy sources, he unfortunately links nuclear energy with nuclear weapons and possible risks for the environment (though advocating comparing risks and benefits).[163] Nuclear energy's benefits are not mentioned or implied.

Still, it is a valuable document and widely welcomed, though it received a critical Christian response.[164] Two politicians noted: "the encyclical makes only a passing and rather negative reference to nuclear energy. All serious estimates of how a substantially decarbonised world economy can be achieved require a substantial contribution from nuclear energy."[165]

---

161. SRTP, "What Future for Nuclear Power?" The citation is located in the subsection, "A balance of risks and impacts suggests a continuing role for nuclear power."

162. Pope Francis, *Laudato Si'*, 5. Note that the UK's national academy of science, the Royal Society prepared a FAQ (frequently asked questions) sheet for bishops and communities to use in response to questions on science and climate change that may arise in response to the Pope's encyclical. In 2016 the Pope spoke about destroying the environment as a sin. See McKenna, "Pope Francis"; and Pope Francis, "Show Mercy to Our Common Home."

163. Pope Francis, *Laudato Si'*, 30, para. 104 and 184.

164. The Church of England welcomed the Pope's encyclical: Church of England, "Church of England Welcomes." Also, see Wiseman, "In Support." Here Jennifer Wiseman commended it on behalf of the American Association for the Advancement of Science (AAAS) in its program Dialogue on Science, Ethics, and Religion (DoSER). AAAS represents 10 million scientists globally. See also the valuable contribution by Yale University authors Tucker and Grim, "Integrating Ecology and Justice."

165. Forster and Donoughue, *Papal Encyclical*, 6. They are members of the House of

The authors, Peter Forster and Bernard Donoughue, reasonably ask about the Pope's and the Catholic Church's view on nuclear energy since it is not evident. Their views, in turn, have been criticized by David Atkinson for seriously misrepresenting the Pope's position. However, Atkinson agrees with their assessment of the Pope's position on nuclear energy. Indeed, he adds:

> My own view (shared by the UK All-Party Parliamentary Group on Thorium Energy) is that thorium is a much safer nuclear fuel than uranium, and with significantly less polluting waste. Why is not the UK, like several other countries and supported by CERN, investing heavily in thorium research?[166]

This represents an endorsement of research into nuclear fuel. My view, too, is that thorium fuel merits ongoing research (see chapter 7).[167]

One can equally ask about the views from other churches. Are these clear?

World Council of Churches (WCC)

As cited above, Forster and Donoughue criticize the encyclical's view on nuclear energy, and then they immediately add: "The World Council of Churches has formally rejected the future use of both nuclear energy and fossil fuels, thus guaranteeing both low growth and blackouts."[168] The WCC's rejection of nuclear energy and fossil fuels was evident in its "Statement towards a Nuclear-Free World" (2014), which unfortunately treats nuclear weapons and civil nuclear power together without distinction. Canadian Patrick Moore (former Director of Greenpeace International) admits to making this mistake with colleagues when he was an anti-nuclear campaigner: "*We made the mistake of lumping nuclear energy in with nuclear weapons, as if all things nuclear were evil. I think that's as big a mistake as if you lumped nuclear*

Lords. Peter Forster is the Anglican Bishop of Chester while Bernard Donoughue is a Labour peer and lay Catholic. Forster and Donoughue's response was published by the Global Warming Policy Foundation (GWPF). Both are trustees.

166. Atkinson, "*Well-Meaning but Somewhat Naïve*"?, 7. David Atkinson was Bishop of Thetford (2001–2009). Previously, he served on Operation Noah's board. For the All Party Parliamentary Group on Thorium Energy, see http://www.appg-thorium.org.uk and Endicott, *Thorium-Fuelled Molten Salt Reactors*.

167. The Birmingham Policy Commission, *Future of Nuclear Energy*, 39, supports thorium research in the UK.

168. Forster and Donoughue, *Papal Encyclical*, 6.

medicine in with nuclear weapons."[169] It is a pity that the WCC's statement replicates this error. The Church of Scotland addresses them separately.[170]

WCC's negative nuclear narrative regrettably omits the benefits of civil nuclear energy. It states: "By fuelling our economies with nuclear power and protecting ourselves with nuclear weapons, we are poisoning the earth and generating risks for ourselves, our descendants and other living things." But the statement is silent on the damage caused by fossil fuels to the planet and life. It also lacks any balanced assessment of pros and cons in electricity generation.

Moreover, a 2016 economic assessment of the feasibility of replacing nuclear power in South Korea by renewables shows it is unachievable, being based on perceived benefits and costs and customers' "willingness to pay."[171] Interestingly, the authors found that people living near nuclear plants had negative feelings towards renewables but those with negative perceptions of nuclear preferred renewables! They concluded: "people have opposing views regarding nuclear power and renewable energy."[172]

Sir John Houghton and Ian Hore-Lacy

Sir John Houghton is a British evangelical Christian with impeccable qualifications and considerable leadership experience in climate change. He was Professor of Atmospheric Physics at Oxford University and then Chief Executive of the UK's Met Office. He was a founding member of the Intergovernmental Panel on Climate Change (IPCC) and chaired or cochaired its Science Working Group. As a scientist he has constantly drawn attention to the reality and risk of climate change in his writings, meetings, and lectures, influencing many people and groups to wake up and take action, including the US National Association for Evangelicals.[173] I have benefitted from hearing him speak at conferences and reading his papers, articles,

---

169. MacKay, *Sustainable Energy*, 161. Italics original. Moore is an environmentalist who holds a PhD in ecology from the University of British Columbia. See Moore, "Should We Celebrate Carbon Dioxide?" Annual GWPF Lecture, London, October 14, 2015.

170. SRTP, "What Future for Nuclear Power? and "Nuclear Weapons."

171. Park et al., "Can Renewable Energy Replace?"

172. Ibid., 568.

173. His autobiography, *In the Eye of the Storm* (written with Gill Tavner), is an informative read on his work as a scientist and Christian.

and books. He is also founder and former chairman of the UK's John Ray Initiative (JRI), which has published many of his general papers.[174]

He cautiously considered nuclear power in a 2009 article as a non-fossil fuel source to reduce carbon emissions, yet on energy security he asked: "How safe are gas pipelines crossing whole continents? How safe are nuclear power stations from terrorist attack or nuclear material from proliferation to terrorist groups?"[175] These questions made him hesitate on a large nuclear expansion. But they are good questions. However, in a 2010 article he acknowledged that the UK's plutonium (hundreds of tons) could generate electricity in nuclear power stations and thus reduce greenhouse gas emissions while degrading the plutonium (reducing proliferation risks).[176] His 2013 autobiography reiterated this point and favored life extensions of nuclear plants while still seeing disadvantages in nuclear's costs and possible proliferation.[177]

His book *Global Warming* supports nuclear. Again, he raises questions on energy security for gas and nuclear.[178] Then he addresses nuclear's benefits:

- no greenhouse gas emissions (apart from a small amount in construction);
- uses up fuel slowly compared to the total resources available;
- efficient in large units to supply the grid and large urban communities;
- the construction technology is now familiar, so savings in CO_2 emissions can be made;
- recent estimates show nuclear electricity costs are similar to natural gas when carbon capture and storage costs are added to gas;
- the International Energy Agency scenarios assume substantial nuclear growth.[179]

174. JRI is an UK evangelical charity connecting the environment, science, and Christianity. It publishes papers, runs conferences, and offers a Christian Rural and Environmental Studies distance learning course. The website (www.jri.org.uk) provides valuable information. It is named after the English naturalist and Christian theologian John Ray (1627–1705).

175. Houghton, "Sustainable Climate," 31. See also Houghton, *Global Warming, Climate Change*, 9, where he cites Professors Socolow and Pacala (Princeton University) "Stabilization Wedges," on how nuclear power could help reduce emissions.

176. Houghton, "Changing Global Climate," 211. He draws on Wilkinson, "Management of the UK Plutonium."

177. Houghton with Tavner, *In the Eye of the Storm*, 277. Later (279) he recognizes that Lynas, *Nuclear 2.0*, makes a strong case.

178. Houghton, *Global Warming* (5th ed., 2015), 301.

179. Ibid., 313–14.

Of course, Sir John recognizes concerns regarding waste, accidents, and proliferation. He mentions that the new Generation IV reactors address these issues. Advanced reactors should contribute to future electricity supply. Finally, he recognizes the potential for nuclear fusion. In an earlier message to the Church he states:

> we live in a world which is God's world, it's his creation, he's made it and he's told us to look after it. And allowing climate change because of human activity to continue at the current rate is something that is completely against what God has told us to do.[180]

Ian Hore-Lacy is an Australian Christian and scientist who works in the nuclear industry as an educator. I heard him (then Head of Communications of the World Nuclear Association) speak on "The World Nuclear Energy Picture" and in a seminar group on "Stewardship of God's Creation."[181] His book *Responsible Dominion* offers a Christian approach for us as stewards. He includes a balanced examination of energy supply options (see his chapter 5), endorsing nuclear energy.[182] Hore-Lacy challenges some of the romantic approaches to the environment taken by green Christians. Nuclear technology can replace fossil fuels to help mitigate climate change. He is currently Senior Research Analyst for the World Nuclear Association (based in London) and author of the book *Nuclear Energy in the 21st Century*, a clearly illustrated introduction.

Christian Charities and Mission Societies

Sophie Harding's Tearfund booklet *For Tomorrow Too* notes that nuclear is: "viable, readily available, and it creates minimal CO_2—yet nuclear power remains a political hot potato."[183] She observes that environmentalists hold different positions on costs, benefits, and waste. Sensibly, Harding recognizes any energy solution has pros and cons which decision makers must consider.

Christian Aid opposes fracking but is more open to nuclear as "a low-carbon alternative to fossil fuels and thinks that it could have a role to play

---

180. Houghton, "Q&A: Sir John Houghton," 5. Interview by Jonathan Langley.

181. At a day conference titled "Energising the Future," at Redcliffe College, Gloucester, in 2011.

182. See also Hore-Lacy's interview about his book and his Christian views, broadcast for Radio National's *The Religion Report*, September 13, 2006.

183. Harding, *For Tomorrow Too*, 12.

in transitioning to a clean world."[184] Nevertheless, nuclear waste's long-term impact on the environment and communities concerns them, so they prefer renewables as sustainable, with reduced potential adverse consequences. Al Roxburgh (Christian Aid's campaign manager, and a Christian) addressed the issue of a society moving away from dirty fossil fuels, but he omitted any mention of the current or future role of nuclear energy.[185]

Christian Ecology Link's (CEL) 2006 report *Faith and Power* called for more efficient energy usage, conservation, and more renewables, yet rejected nuclear power. Its author, Tony Cooper, stated that reducing harm from climate change is important and he presented an energy strategy which is clear and commendable: "we seek an energy strategy that reflects love of the Creator, expresses care for the whole creation, and is informed by Christian principles of wise stewardship, peacemaking, justice, loving our neighbours and moderation in consumption."[186] Now, I agree with his Christian principles with calls for honesty and transparency, but whereas he argued for a non-nuclear future I take a more positive approach for a nuclear future.[187]

An issue of BMS World Mission's *Mission Catalyst* addressed climate change.[188] Actually, David Kerrigan's editorial summarizes the reports of the Intergovernmental Panel on Climate Change (IPCC) and he asserts the need for alternatives to fossil fuels, mentioning nuclear as one option with "relatively low-carbon." Cutting fossil fuels "and moving towards a lighter carbon-footprint is the ethically responsible option for us all."[189] Another article saw nuclear energy having a role as a reliable, well-understood technology with a future in delivering heat, and using thorium fuel instead of uranium.[190]

---

184. Christian Aid, *Big Shift Q&A*.

185. Roxburgh, "Moving from Gloom to Action," writing in *Mission Catalyst*, 2014.

186. Cooper, *Faith and Power*, 1.

187. Cooper, *Green Christianity*, 200–205, also addresses his concerns over nuclear energy.

188. "Climate Change Is Boring," issue 3, 2014.

189. Kerrigan, "Bored to Death," 2. David Kerrigan is General Director and Managing Editor.

190. Boston, "Can Science Save?," 12–13. Boston previously worked for E.ON on energy systems. At the time of writing he led a team for Energy Research Partnership (working with government, industry, and academia).

United States of America

The scientist and Christian Sir John Houghton (see above) has travelled to the US and engaged American Christian leaders, particularly those in the National Association of Evangelicals (NAE), on the importance of climate science and the threats of global warning.[191] He describes the process by which the NAE accepted the importance of biblical care for creation.[192] Today, Christian leaders such as climate scientist Katherine Hayhoe, at Texas Tech University, continue to communicate climate change.[193]

But where do US Christians stand on nuclear energy? Interfaith Power & Light helpfully lists religious statements of climate change for various faiths with links. It "works to promote ethical, moral, just and sound solutions to global warming."[194] But its position is to reject nuclear energy on the following grounds: cost, timeline, safety, and justice. Rather, it focuses on energy efficiency, conservation, and renewables.

In 2008 the Presbyterian Church (USA) published *The Power to Change*, which addressed nuclear energy.[195] Among its recommendations it included "a moratorium on all new coal-fired and nuclear power plants until related environmental concerns are addressed." It expressed concerns on nuclear proliferation, waste disposal, and safety, but considered it a bridging power to "a more sustainable future," noting: "If the waste and related safety issues cannot be resolved with a very high degree of confidence and integrity, nuclear power should be phased out."[196]

But the Cornwall Alliance has a different perspective. The Cornwall Alliance is an American evangelical Christian organization using biblical principles to promote "environmental stewardship and economic development." The Alliance includes "theologians, pastors, ministry leaders,

---

191. NAE's mission "is to honor God by connecting and representing evangelical Christians." It is influential with 45,000 local churches, about 40 different denominations, and serving millions of people. NAE, "About NAE."

192. Houghton with Tavner, *In the Eye of the Storm*, ch. 23, "America the Potentially Great." However, his experience with the Cornwall Alliance for the Stewardship of Creation was less straightforward.

193. Hayhoe frames her message in terms of the "'Three Rs' of climate change: Reality, Risk and Response" (Hayhoe, "Biography"). See also her 2016 interview on *100 Huntley Street*, Hayhoe, "Truth about Climate Change"; and her lecture, Hayhoe, "Climate Change: Facts, Fictions."

194. Interfaith Power & Light, "Nuclear."

195. ACSWP, *Power to Change*, 4.

196. Ibid., 17.

scientists, economists, policy experts, and committed laymen."[197] They argue for continued use of fossil and nuclear fuels while standing against "harmful climate change policies." "An Evangelical Declaration on Global Warming" (2009) explains their position.

In "An Open Letter to Pope Francis on Climate Change" (October 2016) they respectfully raise "matters of concern," e.g., worldviews, science, and the poor. It is endorsed by over 500 people, including many academics. One section, "For the Foreseeable Future, Wind and Solar Energy Cannot Effectively Replace Fossil Fuel and Nuclear Energy," summarizes the merits of fossil fuels and endorses nuclear energy.[198] A detailed defense for ongoing use of fossil and nuclear fuels occurs in *A Call to Truth, Prudence, and Protection of the Poor 2014*.[199] This document was sent with an open letter explaining how the Alliance differs from other evangelicals. The letter concluded: "We recognize that reasonable people can disagree with our understanding of the science and economics. But this is indeed our understanding."[200] I agree with this stance on nuclear energy.

Environment, Energy, and Ethics

Last, but not least, it is essential to address the environment, energy, and ethics. Many organizations have statements on ethics, e.g., the UK's Engineering Council and Royal Academy of Engineering has its 2014 *Statement of Ethical Principles for the Engineering Profession*. Further, the Chartered Management Institute (CMI) has two 2014 publications: *Managers and the Moral Maze* and *Managers and Their MoralDNA*. The second states:

> Unethical behaviour hurts people in every walk of life. Hospital patients suffer and die, customers are ripped off, investors see their savings destroyed and people lose their jobs, their homes and their self-respect.

---

197. Cornwall Alliance, "About."
198. Cornwall Alliance, "Open Letter to Pope Francis."
199. Legates and van Kooten, *Call to Truth*.
200. Cornwall Alliance, "Open Letter to the Signers." There is a detailed survey of American views on global warming by Roser-Renouf et al., *Faith, Morality and the Environment*, from Yale University and George Mason University, 2016.

> We do not face a crisis of politics, of religion or of economics; we face a crisis of ethics. And our responses to wrong-doing are not only failing, they are making matters worse.[201]

Now we don't have space to follow this up here, apart from noting that the report identifies that religious managers scored higher than non-religious managers in the values and ethical areas considered.[202]

The relevance of ethics in climate change was addressed by the Intergovernmental Panel on Climate Change (IPCC). The IPCC notes the role of ethics and economics within the United Nations Framework Convention on Climate Change (UNFCCC), since Article 2 aims for "avoiding dangerous anthropogenic interference with the climate system." But judging what is dangerous includes ethics and value judgements.[203] The role of nuclear energy in electricity generation is a clear component of this global dialogue on climate change, so let us turn to this.

When I worked as a tutor in the nuclear industry (and later as a consultant) I taught staff about ionizing radiation, e.g., sources, effects, the system of radiation protection, discharges, and risks. We considered international recommendations, legislation, company safety rules, guidance, and controls. For example, we considered radiation doses and risks involving cancers, cataracts, and genetic effects. Also, we compared the average annual risk of death from non-nuclear accidents in the UK with cancer risks from radiation exposure in employment. Moreover, we looked at death in the UK from some common causes (e.g., accidents on the roads) compared with public doses and risks received from nuclear discharges. Staff were trained to know what the risks were for themselves and the public (which of course included their families, friends, and neighbors).

More recently, an article in the *Journal of Radiological Protection* asked about protecting the most sensitive people.[204] It sensibly mentions that the ICRP's 2007 international recommendations "are based on a population average, rather than on the available data for subpopulations. From an ethical

---

201. Steare et al., *Managers and Their MoralDNA*, 6. The International Atomic Energy Agency has its own ethical standards for staff—see IAEA, "IAEA Values."

202. Steare et al., *Managers and Their MoralDNA*, 15.

203. IPCC, *Climate Change 2014*, ch. 3 ("Social, Economics, and Ethical Concepts and Methods"), 211–15.

204. Hansson, "Should We Protect?" Sven Ove Hansson works in the Department of Philosophy and the History of Technology, Royal Institute of Technology, Stockholm.

point of view, this approach is far from unproblematic."²⁰⁵ The article argues that discussing averages is insufficient and such discussions should recognize sensitive groups and provide "the best possible information" to people.²⁰⁶ I agree! Indeed, the international radiation protection bodies are reviewing their ethics. And the World Nuclear Association has a "Charter on Ethics" that member organizations are required to affirm.²⁰⁷

Christian philosopher Michael Northcott noted on the word "ethics": "Its meanings include not just a set of beliefs or values but a way of life, an orientation to the world." Moreover, Northcott stated: "For Christians love, not justice is the ordering virtue."²⁰⁸ Christians authors have made important contributions on ethics.²⁰⁹ Sadly, space limits discussion, but I specifically wish to mention two Christian contributions on nuclear energy.

Christian authors Andy Mellen and Neil Hollow, in the context of "peak oil," reviewed the UK's energy options in *No Oil in the Lamp*, challenging individual readers ("What Can I Do?") and churches ("What Can My Church Do?") on their attitudes and roles. A considerable merit of their work is that they examine the pros and cons of energy options and raise moral issues. Their chapter 3 asks: "Can't we just make do with coal, gas or nuclear power?" They wonder if nuclear can make a useful future contribution to our energy needs, but state: "We think nuclear power raises a number of very tricky moral and ethical issues that Christians rarely consider, and we examine these . . ."²¹⁰ Starting at the key issue of uranium supply, they raise concerns (e.g., past performance, safety, accidents, waste, and the link with nuclear weapons) which I address in subsequent chapters. Their overall assessment is largely negative on nuclear.

---

205. Ibid., 211. International recommendations are produced by the recognized International Commission on Radiological Protection (ICRP). These recommendations are made into law in countries. For example, based on the ICRP recommendations the EU issues directives which are made into UK legislation.

206. Ibid., 217. For example, Hansson mentions that sensitive groups for cancer risk include: women aged 18–64 (compared with men), young children, and the unborn. Life expectancy and cancer survival rates for developed/industrialized countries and undeveloped countries are also discussed.

207. WNA, "Charter of Ethics."

208. Northcott, "Sustaining Ethical Life," 230.

209. For example: Gushee, "Environmental Ethics;"; Moo, "Eschatology and Environmental Ethics"; Hodson, Margot, *Uncovering Isaiah's Environmental Ethics*; Hodson and Hodson, *Ethics of Climate Skepticism*.

210. Mellen and Hollow, *No Oil*, 40.

A 2013 paper by two Christians, Victor Christianto and Florentin Smarandache, at the University of New Mexico, rejects nuclear energy: "Based on ethical, ecological, and economics considerations, the writers conclude that it is not acceptable to consider nuclear energy as part of national's [sic] energy mix, especially for developing countries."[211] They offer reasons for choosing just renewables along with coal and gas as a short-term bridging fix. I agree with some of their points, but not all. My chapters 5 and 6 address some of their concerns.

Finally, the Forum on Religion and Ecology at Yale University had a 2015 news article outlining important initiatives in this area.[212] In particular, five influential US theologians worked collaboratively on a widely reviewed article, "Catholic Moral Traditions and Energy Ethics for the Twenty-First Century," which considers nuclear energy.[213] Concerns and questions on nuclear are, of course, raised and discussed but there is no outright rejection of current nuclear energy. Rather, it is regarded as a limited "bridge fuel" in the present between energy of the past (fossil fuels) and energy of the future (renewables).[214] Nevertheless, they express a desire to restrict further nuclear capacity until an adequate solution is found to the storage of used fuel. At the same time, they doubt whether nuclear can be expanded on sufficient scale to reduce CO_2 emissions by 2030.[215]

It is worth mentioning that the International Commission on Radiological Protection (ICRP) has run international conferences and workshops in Asia, Europe, and America addressing the ethics of radiation protection.[216] Academics have made useful contributions here. For example, in 2014 at the 2nd International Symposium on the Ethics of Environmental Health, Stephen Gardiner contributed a challenging paper on nuclear ethics, outlining new principles in the debate.[217] It is worth noting

---

211. Christianto and Smarandache, *Christian Ethics Consideration*, 8. Christianto studied engineering, gravitation, and cosmology, and holds a Doctor of Divinity. Smarandache is a mathematician.

212. Patenaude, "On the Ethics of Energy."

213. Lothes et al., "Catholic Moral Traditions," 18–23. This article updates a 1981 document by the bishops on "Reflections on the Energy Crisis," looking at the US situation.

214. Ibid., 10–15, 15–23, 23–28. Natural gas is also seen as a bridge fuel.

215. Ibid., 22.

216. See ICRP, "Workshops on the Ethics of the System of Radiological Protection: Supporting Task Group 94." Papers from the workshops are available to download at http://www.icrp.org/page.asp?id=237.

217. Gardiner, *Need for a Public "Explosion"*. He is Professor of Philosophy and Ben

that radiation protection and ICRP recommendations are not restricted to safety in nuclear plants; they are much wider in scope.

Although it is beyond the scope of this book, which is focused on electricity production, there are around 245 research reactors worldwide and another 180 nuclear reactors used for powering ships and submarines.[218]

Conclusion

As I worked in the nuclear industry for many years, I provide a former insider's perspective. I favor nuclear energy, but avoid acrimonious arguments regarding nuclear versus renewables. My view is we need both to avoid catastrophic climate change (plus other actions, e.g., reduced consumption). Nuclear provides clean and secure energy. Some environmentalists now support it and have exchanged their former negative nuclear narrative for a more positive narrative. I have covered a range of views but readers may have noticed that I have omitted the general public's attitudes and understanding. These are examined next. I accept genuine concerns and questions so my next chapters address particular issues. Importantly, churches and Christians need to engage in respectful dialogue on how we use energy technologies wisely for our lives, our neighbors, other creatures, and our planet. It is God's world and we should care for it. This approach encompasses Christian mission worldwide.

Rabinowitz Endowed Professor of the Human Dimensions of the Environment at the University of Washington. Also, Taebi, *Good Governance of Risky Technology* (presented at the ICRP's Second North American Workshop, 2015) helpfully addressed values, norms, and design criteria. He noted the importance for risky technology of considering the aspects of social acceptance and ethical acceptability. See Taebi and Roeser, "Ethics of Nuclear Energy."

218. WNA, "Research Reactors" and "Nuclear-Powered Ships."

Further Reading

Environment Agency, et al. *New Nuclear Power Stations: GDA*. Infographic March 31, 2017. https://www.gov.uk/government/uploads/system/uploads/attachment_data/file/605324/GDA.pdf.

Houghton, John. "Q&A: Sir John Houghton." Interview by Jonathan Langley. *Mission Catalyst* 3 (2014) 4–5. http://www.bmsworldmission.org/engagecatalyst/mission-catalyst-climate-change-is-boring/qa-sir-john-houghton.

International Energy Agency. "IEA Executive Director Delivers Keynote Address at World Nuclear Exhibition." June 28, 2016. http://www.iea.org/newsroom/news/2016/june/iea-executive-director-delivers-keynote-address-at-world-nuclear-exhibition.html.

Irvine, Maxwell. *Nuclear Power: A Very Short Introduction*. Oxford: Oxford University Press, 2011.

Nuclear Energy Institute. *Nuclear Energy: Powering America's Future*. October 2016. https://www.nei.org/CorporateSite/media/filefolder/Publications-Brochures/Brochures/2016_New_Plant_Brochure.pdf.

World Nuclear Association. *Nuclear Power Reactor Characteristics*. 2016/17 Pocket Guide. August 2016. http://www.world-nuclear.org/our-association/publications/pocket-guides/pocket-guide-nuclear-power-reactor-characteristics.aspx.

Yeo, Tim. *Five-Point Plan for the UK Nuclear Energy Industry*. New Nuclear Watch Europe, February 27, 2017. http://newnuclearwatch.eu/five-point-plan-for-the-uk-nuclear-energy-industry/.

Questions

1. Read David Atkinson's *Climate Change and the Gospel: Why We in the Churches Need to Treat Climate Change More Urgently* (http://operationnoah.org/wp-content/uploads/2015/02/Climate-and-Gospel-David-Atkinson-30-01-2015.pdf). He begins: "Why are we in the Christian Church so slow in responding to the most important moral issue of our generation: climate change?" How do our views on nuclear energy impact on this critical challenge?

2. Use the World Nuclear Association's website (http://www.world-nuclear.org) to search for a country of interest to find out their nuclear energy situation, e.g., the United Kingdom, United States, France, or Germany.

3. Watch Professor Sir David King's lecture on "Renewable Energy and Nuclear Power for the UK" (https://www.youtube.com/watch?v=nPQFQDIg3QM). How does his lecture help you to understand nuclear energy's role?

4. In May 2017 the UK's political parties issued their manifestos for the June 2017 general election. The manifestos of the three main political parties (Conservatives, Labour, and Liberal Democrats) were summarized in *World Nuclear News* ("UK Parties Make Scant Reference to Nuclear Power." May 19, 2017. http://www.world-nuclear-news.org/NP-UK-parties-make-scant-reference-to-nuclear-power-19051701.html.). Either use the article here or look up their manifestos online to find out their views on nuclear energy in the UK. For other parties' views look these up online, e.g., Green Party (https://www.greenparty.org.uk/), Plaid Cymru (http://www2.partyof.wales/), Scottish National Party (https://www.snp.org/), and UKIP (http://www.ukip.org/). Are they silent, supportive, or something else, on the UK's current nuclear plants and any future new nuclear plants?

5

Nuclear Energy
Concerns, Questions, Objections (Part 1)

> ...industry leaders acknowledged that many in the public now felt more fear than respect, more anger than awe. Electricity had proved to be a powerful but "dangerous servant," and some wondered aloud if it was doing more harm than good. Newspapers admonished the electric companies for unleashing a destructive force that they seemed unable to control.
>
> ERNEST FREEBERG, ON THE 1890 CONVENTION OF
> THE NATIONAL ELECTRIC LIGHT ASSOCIATION[1]

> Please don't get me wrong: I'm not trying to be pro-nuclear.
> I'm just pro-arithmetic.
>
> DAVID MACKAY[2]

---

1. Freeberg, *Age of Edison*, 179. The 1890 convention of the National Electric Light Association in Kansas City, US. Freeberg noted the attendees' "faith in their industry's future had been shaken by a recent series of electrical fires and accidents that had provoked a public backlash." After the mayor's welcome they focused on public concerns to this new technology called electricity.

2. MacKay, *Sustainable Energy*, 169. He is commenting on his calculation of the carbon intensity needed to construct a nuclear power plant. Sir David MacKay was a British physicist and academic. His positions included Regius Professor of Engineering at the University of Cambridge and Chief Scientific Advisor to the Department of Energy and Climate Change.

NUCLEAR ENERGY: CONCERNS, QUESTIONS, OBJECTIONS (PART 1)

Introduction

THIS CHAPTER CONSIDERS THE crucial position of public acceptance and understanding of nuclear power, including five objections. Remember, this is an introductory book. My responses to issues are not comprehensive; rather, they aim to introduce answers to start readers thinking. For more detail consult my references. Let us begin!

Public Attitudes and Understanding

The first quotation above illustrates fears expressed among the public and the media in late-nineteenth-century America over this new stuff called electricity. Attendees at the convention had to address such fears. Today, public attitudes and understanding of nuclear energy are also important and need addressing. University of Bristol Cabot Institute scientists and volunteers asked the public at Bristol's Harbour Festival (July 2015): "Is Nuclear Green?" Their aim was to encourage public debate and discussion on nuclear energy in the UK, challenge nuclear myths, present relevant science, and consider principles related to radioactive waste.[3] The public engaged in discussions and had the option of polling for nuclear as a green energy source or not. This university event was supported by nuclear industry employees from Magnox Ltd. and EDF Energy, so they allowed people opportunities to question them. Participants recognized that often nuclear energy had a "negative public opinion" for three main reasons: waste, accidents, and a historical link with nuclear weapons. Let us consider earlier public attitude surveys.

In 2006 (after Chernobyl but before the 2011 Fukushima accident) Wouter Poortinga and colleagues published *Public Perceptions of Nuclear Power, Climate Change and Energy Options in Britain*. It summarized a 2005 UK survey of attitudes towards electricity generation, nuclear power, and climate change. There were longstanding "negative connotations" among the public on nuclear power and waste but also new concerns over climate change. Its authors admitted: "almost nothing is known about how ordinary citizens will respond to the current policy reframing of these issues around the arguments that link nuclear energy to climate change threats, energy reliability and wider sustainability objectives."[4] Two interesting sections

---

3. University of Bristol Cabot Institute, "Is Nuclear Green?"

4. Poortinga et al., *Public Perceptions*, 1. Professor of Environmental Psychology, Cardiff University, UK.

on "Attitudes Towards Nuclear Power" (section 2) and "Attitudes Towards Climate Change" (section 3), when compared, demonstrated "that climate change is viewed in a far more consistently 'negative' light than nuclear power is."[5] Their investigation showed that the challenges of nuclear energy and climate change pose no simple answers but identified a "reluctant acceptance" of nuclear energy to combat climate change.[6] Indeed, the report early on indicates this approach: "pursuing the nuclear option was judged by many in the [focus] groups as the lesser of two evils. This discourse, of what we have termed a 'reluctant acceptance' was a common response to the explicit reframing of nuclear power in terms of climate change mitigation."[7] In other words, nuclear energy becomes more acceptable when viewed in the context of climate change solutions.

Research published in 2011 by Adam Corner and colleagues further explored "reluctant acceptance," changes in UK public attitudes from 2005 to 2010, and relationships between nuclear, climate change, energy security, and environmental values. They reviewed research on British public attitudes, recognizing that "Public support is likely to be one of the most important factors that will determine future technological pathways that the UK and other societies eventually take in the face of climate change and energy security."[8] Nevertheless, in their detailed analysis they noted that British public opinion on nuclear energy remains divided, while concerns over climate change and energy security may, in limited circumstances, increase nuclear energy's acceptability.

Then in 2010 the OECD's Nuclear Energy Agency (NEA) investigated public attitudes.[9] This report recognized that large numbers of the public hold no firm views on nuclear energy. This "middle group" with their attitudes is crucial to governments. Moreover, the NEA identified a "clear correlation between knowledge and support." Also, the publication stated that when the climate change benefits of nuclear energy were explained support for nuclear rose significantly. Public involvement is crucial for an expansion of nuclear power plants. But three factors reduce public support: terrorism, waste, and "misuse of nuclear materials." Where does the public receive its information

---

5. Ibid., 13.
6. Ibid., 23.
7. Ibid., 2.
8. Corner et al., "Nuclear Power, Climate Change," 4825. In 2011, Adam Corner worked in the School of Psychology, Cardiff University, UK.
9. NEA, *Public Attitudes to Nuclear Power*.

on energy, including nuclear? The report identifies the media, yet noted that the public "does not trust it"! Next, national governments are trusted less than media sources—so governments have a problem in educating/influencing "their publics." Communications must be "open, honest and balanced." The most trusted groups are: scientists, environmental groups, and consumer organizations.[10] Interesting observations, don't you think?

In the US, the Nuclear Energy Institute (NEI) reported from a survey on "US Nuclear Plant Neighbors and Public" (Fall 2015), in which respondents were asked how strongly they supported or opposed nuclear generated electricity in the US (with five options). The results were surprising. Plant neighbors in June 2015 gave 83 percent support (50 strongly favor and 33 favor), while in March 2015 the general public support was 68 percent (27 strongly favor and 42 favor).[11]

Returning to the UK, in October 2015 the UK's Institution of Mechanical Engineers (IMechE) press office released: "New Survey: Majority of the Public Support UK Nuclear Power." The survey covered the public in September 2015 and found: "56% of the public support the UK continuing to use nuclear power, compared with just 19% who do not and 25% who were unsure." Among supporters the reasons were: keeping the lights on (82 percent); job provision (56 percent); and boosting the economy (54 percent). Opposition, though, centered on: danger (77 percent); environmental damage (76 percent); and expense (27 percent). Moreover, 44 percent didn't want a "nuclear waste facility" within 10 miles of their home but 32 percent wouldn't object. Jenifer Baxter, the Institution's Head of Energy and Environment, felt encouraged by the response. Yet, she noted knowledge of nuclear technology and waste was insufficient; industry and Government needed to raise public awareness of benefits and safety.[12]

In the UK, the nuclear industry and government encourage public engagement. For example, the 2012 House of Commons Science and Technology Committee's (HCSTC) *Devil's Bargain? Energy Risks and the Public*

---

10. Ibid., 7–8.

11. NEI, *NEI US Public Opinion*, 1. For further research over a six-year period, see Bisconti Research, *6th Biennial Survey* (Summer 2015). This shows ongoing support for nuclear energy. A more recent poll (September to October 2016) by Bisconti Research, Inc., and Quest Global Research conducted for the NEI on long-term trends, continues to show strong support among Americans for nuclear energy. See NEI, "Public Opinion: Americans Voice Strong Support."

12. IMechE, "New Survey."

includes written evidence from witnesses. They particularly commended the work of the Science Media Centre:

> . . . the Society for Radiological Protection considered it to be "an important objectively-based way of resisting 'quackery' and providing good quality information to the media, should they choose to use it." Professor Pidgeon stated that the SMC "has been a very good development in this country. They have connected the journalists with the scientists and engineers in a very effective way over many years on many issues."[13]

Nevertheless, more work is still required to connect journalists and scientists for improving risk communication in the UK media.

In 2013 the UK's *Nuclear Industrial Strategy* identified the importance of public opinion and engagement and noted the work done by many groups to address public concerns, recognizing: "It is important to continue to help improve the public understanding of risks and of radiation, as well as the benefits of the use of nuclear."[14]

Following this strategy document, the Nuclear Industry Council (NIC) published *In the Public Eye: Nuclear Energy and Society* (2014), stating the need to develop a strategy which will "develop a consistent nuclear narrative highlighting the contribution nuclear energy makes to benefit society."[15] The strategy aims to "maintain and enhance public confidence in nuclear energy." This report shows that over the last 10 to 15 years in the UK public support for nuclear energy has grown, although slowly and with a dip after the Fukushima accident. So it is "critically important" for the government, industry, and the expert community to interact with the public over technology and benefits to society, since public attitudes on nuclear and its acceptability are critical. Drawing upon previous work, they identify that public attitudes to technology formed six different groups (confident engagers, distrustful engagers, late adopters, concerned, disengaged sceptics, and indifferent).[16] Although the report acknowledges that these groups refer to technology in general they are appropriate for nuclear energy. Consequently, such target audiences, they recognize, need to be specifically engaged.

13. HCSTC, *Devil's bargain?*, 27.
14. BIS, *Nuclear Industrial Strategy*, 48.
15. NIC, *In the Public Eye*, iii.
16. Ibid., 4, using Ipsos MORI, *Public Attitudes to Science 2014*. This is an excellent source on much wider aspects of public attitudes with valuable discussions on science, religion, and nuclear energy.

NUCLEAR ENERGY: CONCERNS, QUESTIONS, OBJECTIONS (PART 1)

But what does all this government and industry effort mean in practice for the public? Well, the Nuclear Industry Council's *Nuclear Energy and Society: A Concordat for Public Engagement* (2015) lays down four key principles for engagement: (1) leadership commitment, (2) best practice, (3) effective communicators, and (4) making a difference. Best practice is characterized by four components: dialogue, trust, clarity, and consultation.[17] The Nuclear Industry Association (NIA) website is worth visiting for public information, including a range of short videos (e.g., "Public Opinion of the Nuclear Industry" and "Exploring Female Attitudes towards Nuclear Power"). Let's take another example.

In 2015, as part of the UK's Generic Design Assessment (GDA, see chapter 4), 3KQ (Three Key Questions) produced a public dialogue report for regulators, *New Nuclear Power Stations: Improving Public Involvement in Reactor Design Assessments*.[18] This important report shows constructive dialogue between the public and regulators with lessons learnt (by both). A key aspect was improving people's trust and confidence in regulators' decisions. It's worth reading. One question was:

> What would you want to know more about? The top three issues for respondents were safety (82%), radioactive waste management (78%), and the impact of radioactive discharges on people and the environment (76%). This was followed by security (64%), spent fuel management (59%) and 'other environmental impacts' (50%).[19]

Moving into 2016, the Nuclear Industry Association (NIA) published UK survey results (late 2015) by YouGov. This showed: "more people support replacement new build than oppose," however, "men are more in favour of new build than women". Other advantages were identified, but concerns continue over waste, costs, and safety.[20]

In July 2016 the UK Department for Business, Energy and Industrial Strategy (BEIS) noted on public attitudes:

> At wave 18 [July 2016], 36% supported nuclear energy compared with 22% who were opposed. There was a significant drop in both

---

17. For an overview of the "Concordat" see the news release NIA, "Nuclear Industry Commits."

18. Regulators: Environment Agency, Office for Nuclear Regulation, and Natural Resources Wales. (Northern Ireland and Scotland have their own Environment Agencies.)

19. 3KQ, *New Nuclear Power Stations*, 15–16. The first sentence is bold in the original.

20. NIA, *UK Public Opinion*. Here "replacement new build" means new nuclear power plants being built to replace old plants as they are closed.

strong support and strong opposition to nuclear energy. Both were down to 6% at wave 18 from 10% at wave 17 [April 28, 2016] . . .

Four in ten (40%) selected the neutral option at this question, to indicate that they neither support nor oppose the use of nuclear energy.[21]

Wave 19 addressed surveys in September–October 2016. This tracker commented: "A third supported nuclear energy (33%) at wave 19, down from 36% at wave 18 and 38% at wave 17. A quarter were opposed to nuclear energy (26%), up from 22% at wave 18."[22] Those who neither support of oppose nuclear (or had no opinion) amounted to four in ten, or 41 percent.

However, in February 2017 BEIS published its latest information (Wave 20) of surveys conducted in December 2016. For nuclear energy, "In Wave 20 a third supported nuclear energy (36%), while one in five are opposed (20%)." Moreover, the tracker stated: "the proportion saying they neither support or oppose [is] now at its highest recorded level, 42%."[23]

So surveys show public attitudes are important, but what about public understanding? We can have attitudes without knowledge or understanding. Is focusing exclusively on attitudes wise?

Julian Hamm's presentation at the UK Nuclear Academics Meeting (2014) notes the paucity of research on public *understanding*.[24] He demonstrates misconceptions held by the public on radiation and the necessity of assuming that people hold misconceptions and dealing with them. Positively, misconceptions are attempts to understand the world. But people need to be challenged to replace old models and views with new ones. By the way, his investigations with both science and non-science graduates showed misconceptions in both groups. Education of the public is an explicit aim of the Society for Radiological Protection (SRP) and the Nuclear Institute, both UK charities.

Elsewhere, a 2014 paper by Eun Ok Han and colleagues investigated Korean school children and demonstrated that behavior towards nuclear energy can be changed through education.[25] Similarly, a 2015 paper shows

---

21. BEIS, *Energy and Climate Change: Wave 18*, 5. "Wave" is the word used instead of "survey." This publication was previously issued by the Department of Energy and Climate Change (DECC).

22. BEIS, *Energy and Climate Change: Wave 19*, 6.

23. BEIS, *Energy and Climate Change: Wave 20*, 5.

24. Hamm, "Radiation Misconceptions and Public Fears."

25. Han et al., "Korean Students' Behavioral Change."

that educating doctors and citizens with teaching on radiation and risks can lower radiation fears in Japan after Fukushima.[26]

Let us return to Europe. Interestingly, Switzerland had a referendum on the future of its nuclear power plants in November 2016. The public were asked to vote on a proposal to phase out their nuclear plants under a strict timetable. If accepted, then three nuclear plants would be closed in 2017 and two others in the 2020s. The public referendum rejected the initiative that would limit the operational lives of their nuclear plants. Rather, the referendum demonstrated public confidence in their continued operation. Nevertheless, in May 2017 another referendum decided to ban building new nuclear plants, while still retaining the current operating plants.[27]

Churches and Christians appreciate the important historical role of education and action in their mission to teach the good news of Jesus and make disciples (Matt 28:19–20; Acts 1:8–9). This teaching role gently corrects misunderstandings, builds scriptural knowledge, and challenges worldviews and behavior. Like the wise person in Proverbs, we should seek understanding.

Let us look at some specific concerns and objections to nuclear.

Objection 1:
There Is a Link between Nuclear Energy and Nuclear Weapons

American risk author Paul Slovic, writing on risk perception and nuclear power in 1990, quotes Smith on the origin of fears after the bombings of Hiroshima and Nagasaki (1945): "Nuclear energy was conceived in secret, born in war, and first revealed to the world in horror. No matter how much proponents try to separate the peaceful from the weapons atoms, the connection is firmly embedded in the minds of the public."[28] More recently,

---

26. Kohzaki et al., "What Have We Learned?" Nevertheless, Ho et al., "Risk Perception," 788, surveyed communities in Taiwan five months after the Fukushima accident (2011) and showed under a half of respondents supported a fourth reactor. For an academic discussion on Taiwan, see Shih et al., "Socioeconomic Costs of Replacing Nuclear."

27. NucNet, "Switzerland Rejects Plans." Even though in October 2016 a Swiss Parliament 2050 future strategy had chosen to prevent replacing nuclear power plants. Also, see NucNet, "Referendum Result Shows Public Confidence." On the May 2017 decision see NucNet, "Switzerland Votes to Phase Out." NucNet noted that opponents of the nuclear phaseout warned "that the government's plans to push renewables and energy savings were costly, posed risks to energy security and would not be environmentally friendly."

28. Slovic, "Perception of Risk," 2, quoting K. R. Smith, "Perception of Risks Associated with Nuclear Power," *Energy Environment Monitor* 4.1 (1988) 61–70.

Christian authors Spencer and White remark: "Its association, no matter how informal, with nuclear weaponry lends it a sinister shadow that, rightly or wrongly, colours the whole debate."[29] This link concerns many people, so let us examine it.

First, American historian, journalist, and author Richard Rhodes researched and wrote the important book *The Making of the Atomic Bomb* (1986). For it he received the Pulitzer Prize and other awards. His other books on nuclear history followed. On the US scene he wrote "Introduction: Gwyneth's Pilgrimage" for Gwyneth Cravens's book *Power to Save the World: The Truth about Nuclear Energy*. Here he reports his previous anti-nuclear position, but he confesses that while writing *The Making of the Atomic Bomb*:

> ... I got to know the extraordinary men and women who developed the science of nuclear physics, many Nobel laureates among them, and learned to my surprise that they looked at nuclear energy very differently from their perspective of firsthand knowledge than I did from my perspective of secondhand ignorance.[30]

He admits that after further questions, reading, visits, research, and writing for his book he changed his mind to see nuclear energy positively as a solution to environmental health problems and global warming. Note that he sees nuclear positively although he knows its link with weapons. Graham Farmelo's book *Churchill's Bomb: A Hidden History of Science, War and Politics* (2013) also looks at the same issue and the relationship between Churchill and the US. It is a complex story of scientists, politicians, fears, secrecy, and war.

But on a more personal level, I grew up in postwar Liverpool during the Cold War, when nuclear weapons and fears of annihilation were daily thoughts, especially when US President John F. Kennedy stood up to the Soviet premier Nikita Khrushchev in the Cuban Missile Crisis (October 1962).[31] As a teenager in the 60s I visited a police station in central Liverpool where wall maps marked the extensive destruction and death if Russian nuclear weapons were detonated over us. There was no chance of survival in our street.

In school I learnt about atomic and nuclear physics and radioactivity. At university in my undergraduate years studying physics I developed more

---

29. Spencer and White, *Christianity, Climate Change*, 188.
30. Cravens, *Power to Save*, xi.
31. E. R. May, "John F Kennedy."

NUCLEAR ENERGY: CONCERNS, QUESTIONS, OBJECTIONS (PART 1)

understanding in these topics and my final year elective topics included a unit on nuclear physics. For my dissertation I completed a theoretical computer study on the effects of radiation on cells. Then for my master's degree I built on my undergraduate foundation by studying radiation protection (or health physics). Our lecturers explained atmospheric testing of nuclear weapons, radioactive fall-out, and risks to the world population. What was I to do with my knowledge? I chose a career as a scientist on nuclear power plants working in radiation protection rather than atomic weapons work. I knew about the link but by then, as a follower of Jesus (or committed Christian), I saw that peaceful use of nuclear-generated electricity was a benefit I could contribute to.

What is surprising about the link between nuclear power and nuclear weapons is that Japan, the only nation to suffer such an attack, which produced such devastation, disease, and death, has embraced the peaceful use of nuclear energy. Think about that! Raul Deju and colleagues observe that Japan decided not to develop nuclear weapons but became a signatory to the Nuclear Non-Proliferation Treaty. The country has little in the way of natural minerals and energy sources so nuclear provides a stable electricity supply. And besides its home market it has an international interest in owning nuclear companies and promoting its nuclear power expertise. Japan has also signed the Kyoto Treaty and is therefore committed to reducing greenhouse gas emissions—clean nuclear power helps here.[32] If anyone has reasons to reject nuclear power because of the link with weapons, it is the Japanese nation. Yet they have embrace nuclear power. (The situation following the Fukushima accident is considered in the next chapter, since this is a separate issue to the weapons link.)

In May 2016, President Obama became the first serving US president to visit Hiroshima at the remembrance service for the bombings. He embraced a 71-year-old man who survived. Can't we too move on instead of worrying about a nuclear link? I know that people consider that nuclear got off to such a catastrophic start as a weapon of mass destruction that they consider it virtually irredeemable. But is this going to stop the dialogue we expect in a democracy? I pray not. Certainly, the early church got off

---

32. Deju et al., *Nuclear Is Hot!*, 41–44. When this book was published in 2009, Dr. Raul Deju was Chief Operating Officer of EnergySolutions, Inc. He has been a professor in American universities. For more information on Japan, see WNA, "Nuclear Power in Japan," e.g.: "Its first commercial nuclear power reactor began operating in mid-1966, and nuclear energy has been a national strategic priority since 1973."

to a very bad start after Jesus, the Messiah, was crucified and killed as a criminal. Bad things can develop into good things.

An important text here is the well-known statement that they shall "beat their swords into ploughshares." Wikipedia states: "Swords to ploughshares (or Swords to plowshares) is a concept in which military weapons or technologies are converted for peaceful civilian applications."[33] Its origin is the Old Testament (or Hebrew Bible), in Isaiah 2:3–4: "He [the LORD/Yahweh] will judge between the nations and will settle disputes for many peoples. They will beat their swords into plowshares and their spears into pruning hooks. Nation will not take up sword against nation, nor will they train for war anymore."[34] A similar vision is shared by Joel 3:10 and Micah 4:3.

Weapons used for producing death and destruction are turned into tools for farming, food production, and life. Fears about past links must not distort decisions on energy. As mentioned before, Patrick Moore (former Director of Greenpeace International) states: "We made the mistake of lumping nuclear energy in with nuclear weapons, as if all things nuclear were evil. I think that's as big a mistake as if you lumped nuclear medicine in with nuclear weapons."[35]

Objection 2: Uranium Supplies Are Limited

Another objection considers uranium supplies. Will there be enough uranium to fuel current and new reactors? This fuel issue needs addressing fairly. Some people argue, or imply, that uranium supplies are insufficient. Let us look at these views.

Adam Ma'anit's 2005 *New Internationalist* article remarked:

> A growing number of studies tell us that, if we were to replace outright all fossil-fuel generated electricity with nuclear, there would be enough economically viable uranium to fuel the reactors for only three to four years. After that the nuclear revolution would grind to a sudden and catastrophic halt.[36]

---

33. Wikipedia, "Swords to Ploughshares."

34. For other translations, see https://www.biblegateway.com/verse/en/Isaiah%202%203A4.

35. Cited by MacKay, *Sustainable Energy*, 161.

36. Ma'anit, "Nuclear Is the New Black."

NUCLEAR ENERGY: CONCERNS, QUESTIONS, OBJECTIONS (PART 1)

Dinyar Godrej's book repeats his argument in a section "False Alternatives" objecting to nuclear power as part of the solution for climate change.[37] But is this argument valid?

First, Ma'anit does not identify these "growing numbers of studies." Second, in the real world we are not instantaneously going to replace all fossil fuel power stations with nuclear ones. New nuclear will be part of the future energy mix, but not all of it. Third, he qualifies his argument with "economically viable" but does not state his working assumptions. Now, more than ten years after Ma'anit's statement, nuclear reactors continue to operate using uranium fuel and new reactors come online.

Tom O'Flaherty's 2011 article in *Nuclear Future* considers this critical question: "If Use of Nuclear Power Continues, Will There Be Enough Uranium?"[38] He rightly identifies crucial factors related to this question, e.g., annual consumption of uranium in current reactors, expansion of new reactors, recoverable reserves (known and estimated), and how soon fast-breeder reactors will be in operation.

But O'Flaherty correctly points out an essential element in this discussion: uranium is not the only fuel for reactors. The naturally occurring element thorium, which is widely distributed on Earth, can be used in new designs of reactors. Countries such as India and Russia are advancing research here. These thorium-fueled reactors must be considered as future sources of energy.

However, returning to just our uranium reserves, O'Flaherty estimates that 75–100 years of uranium is available using current Generation III reactors. However, fast-breeder reactors (Generation IV) can "breed" their own fuel and give an efficiency of conversion of around 50 times greater than our current reactors. He states: "The life of proven and estimated uranium reserves is therefore increased from tens of years to many hundreds, if not thousands, of years—a scenario which can surely be deemed sustainable, by any reasonable yardstick."[39]

Again, in 2012 Christian authors Mellen and Hollow argued that uranium is fast running out—a genuine concern. They see only 70 years of supply left at current usage but regard this as "undoubtedly optimistic." They examine various options, e.g., further mining and ore strength, and

---

37. Godrej, *No-Nonsense Guide*, 129–30 (130).

38. O'Flaherty, "If Use of Nuclear Power Continues." Dr. Tom O'Flaherty was CEO of the Radiological Protection Institute of Ireland, 1992–2002.

39. Ibid., 33.

fast reactors. But their overall approach lacks optimism. Moreover, information they use is outdated; more recent data is required.[40]

How can we respond? In 2009 David MacKay examined the issue of "sustainability" of nuclear power plants, rightly reviewing our uranium reserves: for a few decades, or millennia?, he wondered.[41] He calculated reserves in the ground, phosphate deposits, and seawater and river water. (Yes, seawater is naturally radioactive with uranium, etc.!) He concluded: "ocean extraction of uranium would turn today's once-through reactors into a 'sustainable' option—assuming that the uranium reactors can cover the energy cost of the ocean extraction process."[42] However, this is not a commercial option at present.[43] MacKay examined other options, including thorium fuel instead of uranium and fast reactors, which use uranium more efficiently.

The Department of Energy and Climate Change (DECC) reviewed uranium supplies in its 2010 report *The Justification of Practices*, which examined the new build of AP1000 reactors in Britain. As part of the process a consultation was conducted and respondents' comments and concerns were addressed. DECC noted:

> Some respondents said that there is a reliable supply of uranium at a cost which is a small part of generating costs. However, a number of respondents doubted the future availability of uranium, and argued that increased demand worldwide could lead to supplies becoming more difficult and expensive.[44]

In response, the section "Reliability of Uranium Supplies" drew upon a wide range of reliable resources to show that there will be sufficient supplies for the proposed new build program.[45] In particular, the Euratom Supply

---

40. Mellen and Hollow, *No Oil*, 39–46. For example, fig. 2 (p. 43) only considers uranium supply and demand from 1947 to 2004.

41. MacKay, *Sustainable Energy*, 162.

42. Ibid., 164. By "once-through reactors" he means the fuel goes through the reactor once. However, it is possible to recycle fuel and return it to reactors to generate more electricity (see my ch. 7).

43. Nevertheless, research work by Carter Abney and colleagues at Oak Ridge National Laboratory (US) is promising, with adsorbents that can extract dissolved uranium from seawater. See Walli, "ORNL Technique." Moreover, on February 17, 2017, Tom Abate at Stanford University in California reported: "Stanford researchers say extracting uranium from seawater could help nuclear power play a larger role in a carbon-free energy future." See Abate, "Stanford Researchers."

44. DECC, *Justification of Practices*, 40, para. 5.33.

45. Ibid., 41–43, para. 5.40–48.

Agency considered that current demand upon our identified resources can be sustained for about a century.

The latest Intergovernmental Panel on Climate Change (IPCC) Assessment Report (the fifth) explains the uranium concentrations and tons in the Earth's crust and seawater. After briefly considering extraction costs, present sources, and estimated sources to be discovered, the IPCC states: "Present uranium resources are sufficient to fuel existing demand for more than 130 years, and if all conventional uranium occurrences are considered, for more than 250 years."[46] Each category would be doubled if spent (used) fuel was reprocessed and recycled to use the remaining uranium and plutonium (but this depends upon economic competiveness). Reactors that use fast-breeder technology can theoretically increase uranium utilization by a factor of 50 or more while reducing high-level waste.[47] Further, thorium, a metal that could be used as nuclear fuel, is widely distributed and three to four times more abundant than uranium in the Earth's crust (although precise figures are unavailable).

The IAEA is involved in organizing publications, projects, and an international seminar on uranium and the nuclear fuel cycle (2014). It provides support and training on exploration and mining uranium resources.[48] On uranium supplies it records: "In planning the supply of uranium fuel for nuclear power plants, owners and operators require accurate information on uranium resources, production and demand worldwide."[49] Consequently, the IAEA jointly issues with the NEA the important resource known as the "Red Book," officially titled *Uranium 2016: Resources, Production and Demand*.[50] It estimates recoverable uranium resources and costs.[51] The World Nuclear Association has a useful discussion on uranium, with tables and graphs. It notes that from 1975 to 2013 uranium resources showed almost a threefold increase following expenditure on uranium exploration.[52]

---

46. IPCC, *Climate Change 2014*, ch. 7 ("Energy Systems"), 526.

47. Ibid.

48. IAEA, *IAEA Annual Report 2014*, 33–34.

49. Ibid., 33.

50. NEA and IAEA, *Uranium 2016*.

51. The "Red Book" reported that future uranium supply was "more than adequate" but future production capacity was a challenge. For useful summaries, see NucNet, "Global Uranium Supply"; and *WNN*, "Red Book Sees Production Capacity."

52. WNA, "Supply of Uranium."

Earlier, in 2015 the IAEA assessed uranium's availability, stating: "The numbers presented for total identified resources show an increase between the 2011 and 2014 editions, to a total of 7.6 million tonnes of uranium (Mt U). This would be sufficient to meet demand for over 120 years, considering 2012 uranium requirements of 61 600 tonnes."[53] They noted a broader estimate of various sources produced a figure of 15.3 Mt U (and even this does not include all sources of uranium, e.g., extraction from chemical phosphates). Although over time the grade of ore deposits will move from higher to lower, because of other factors this doesn't necessarily mean costs will increase. Moreover, the cost of uranium fuel is a relatively small part of nuclear generation. The IAEA concludes: "In general, the literature supports the view that possible shortages of uranium—and their reflection in increased uranium cost—should not be regarded as an obstacle to nuclear energy's making a significant contribution to the reduction of GHG emissions."[54]

Increased exploration means that further supplies of uranium have been identified. For example, in the March 2015 *IAEA Bulletin* there was a report that Tanzania identified about 60,000 tU which it intended to start mining in 2016.[55]

Finally, the NEA's *Managing Environmental and Health Impacts of Uranium Mining* covers current countries and countries hoping to commence mining. The report "aims to dispel some of the myths, fears and misconceptions about uranium mining by providing an overview of how leading practice mining can significantly reduce all impacts compared to the early strategic period."[56] A case study on Kazakhstan shows that its annual uranium production has accelerated and it is now the world's largest uranium producer. Its mining industry makes substantial positive socio-economic impacts nationally, regionally, and locally.[57]

---

53. IAEA, *Climate Change and Nuclear Power 2015*, 63.

54. Ibid., 64.

55. Gaspar and Albertini, "Water Protection Measures," 28–29.

56. NEA, *Managing Environmental and Health Impacts*, 3. For more on uranium supply, see 15–16. Related to this is a 2016 study on Canada's uranium mines that shows how low their GHG emissions are. See *WNN*, "Low Emissions."

57. NEA, *Managing Environmental and Health Impacts*, 111–13. Kazakhstan's annual uranium production accelerated from 3,700 tU (2004) to over 20,000 tU (2012).

Objection 3: Building Nuclear Power Plants Uses Vast Quantities of Concrete and Steel

Another objection is that building nuclear plants uses vast quantities of concrete and steel, producing carbon dioxide pollution. So fossil fuels, which are bad, are needed to support nuclear energy; therefore nuclear is bad. That's an argument. But is this approach helpful?

David MacKay's "Mythconceptions"[58] examines this objection and calculates the sums for a 1 GWe nuclear power station generating over a 25-year reactor life. He estimates the carbon footprint of the 1 GWe nuclear station as 300,000 tCO_2.[59] But then MacKay shows that the "carbon intensity associated with construction" is much lower than for fossil fuel plants. Actually, his 25-year reactor life is too short, as new nuclear plants typically have a 60-year lifespan (possibly longer with life extensions). So using this figure reduces MacKay's figure further. He cites the IPCC value, which also includes fuel processing and decommissioning. This value is much less (one tenth) than for fossil fuel power plants. MacKay remarks: "Please don't get me wrong: I'm not trying to be pro-nuclear. I'm just pro-arithmetic."[60]

Of course, we should point out the obvious to objectors. Building fossil fuel power stations also uses fossil fuel in producing the steel and concrete needed in construction. Even wind turbines use fossil fuel in the concrete and steel needed for construction, and Steve Kidd reckoned they use more steel and concrete than nuclear plants per unit of electricity (kilowatt-hour) produced.[61] Finally, the World Nuclear Association compared the concrete and steel used in the UK's Sizewell B PWR with a modern AP1000 PWR and noted three advantages: "Firstly, the AP1000 footprint is very much smaller—about one-quarter the size, secondly the concrete and steel requirements are lower by a factor of five, and thirdly it has modular construction."[62]

So this objection against construction of new nuclear can be answered. But perhaps we can look forward to the day when clean energy produces all our steel and concrete?

---

58. In MacKay, *Sustainable Energy*, 169ff.

59. tCO2 means tonnes of carbon dioxide.

60. MacKay, *Sustainable Energy*, 169. This remark went viral.

61. Kidd, "Is Climate Change." He is an independent nuclear consultant and economist who previously held senior positions in the World Nuclear Association.

62. WNA, "Advanced Nuclear Power Reactors." The actual amounts used are stated in the article.

Objection 4: Nuclear Power Plants Cannot Be Built Fast Enough to Meet Our Needs

Dinyar Godrej's objections to nuclear power include: "In order to switch to nuclear the world would need to build 3,000 reactors by 2100."[63] She does not tell us where that figure comes from, but remarks that it won't happen for various reasons. And she may be right, but . . .

Again, this is another argument tackled by MacKay, who sees the objection of not building new nuclear reactors fast enough as exaggerated by the use of misleading arguments—a technique he calls "the magic playing field." As an example, he cites the *Guardian* newspaper's environment editor, who summarized a report written by the Oxford Research Group. The editor stated that to make an important impact 3,000 new reactors will need to be built over the next 60 years (approximately one per week every year). Then he stated that the highest building rate achieved hitherto was 3.4 reactors per year. So clearly, to the unwary reader, it looks to be an impossible task.[64] But is the argument valid?

MacKay, however, notes that the comparison is switched during the presentation and so he offers a "more honest presentation," highlighting two errors. First, MacKay notes there is a regional switch from 3,000 new reactors built over 60 years *worldwide* to just *one country* (France). Second, the timescale is switched from over 60 *years* to just *one year*: 3.4 reactors per year is just the building rate for France, not the whole world! On the contrary, MacKay shows that historically the peak nuclear plant building rate (in 1984) was actually 30 GWe per year worldwide, i.e., 30 new plants of 1 GWe in that year. Not 3.4 plants per year—a big difference! So, in summary, one reactor per week (50 per year) is comparable with the 30 per year in 1984. MacKay concludes that the build rate is high but it resides "in the same ballpark" as peak historical build rates. His argument convinces.[65]

Let us add one thought. The UK nuclear industry spent years alerting the government etc., that older reactors (i.e., the first-generation Magnox reactors) were approaching their end of life and urgent action was needed to replace them due to (a) the slow approval process and (b) the length of construction times.[66] Sadly, urgency on energy policy was lacking.

63. Godrej, *No-Nonsense Guide*, 130. See my Objection 2 discussion on uranium.

64. MacKay, *Sustainable Energy*, 171.

65. Ibid. In 2012, Brook, "Could Nuclear Fission," 6–7, developed MacKay's argument to show that rapid large-scale building of nuclear plants is feasible on historical grounds.

66. In fact, the last Magnox reactor at Wylfa, on Anglesey in North Wales, ceased

Mellen and Hollow remarked on the industry's record for building nuclear: "no power stations have been built to time or cost or original design, probably anywhere, but certainly in the UK and the US."[67] No evidence is provided to support the statement. Actually, it has been a very long time since the last UK reactor, Sizewell B, began construction in 1988 and started generation in 1995. But is the statement entirely true? Well, not for Sizewell B. Nuttall and Earp's careful analysis of "Nuclear Energy in the UK" stated: "the plant was built to time and cost."[68]

Moreover, the UK's Committee on Climate Change (CCC) remarked: "Nuclear plants have been delivered to cost internationally (e.g., in China and Korea) but have suffered delays and cost overruns in Europe and the United States," and "In recent years, nuclear plants have been delivered to cost and on time in Asia, but have experienced delays, and cost overruns in both Europe and the US."[69] So Asian countries *have* built nuclear plants to time and cost. Perhaps we can learn from them?

Some object to the long timescale for building new reactors, but is this unique? Again, the Committee on Climate Change explains:

> Investments in the power sector have long lead-times, with planning cycles stretching well beyond the current 2020 policy window. Large offshore wind farms, CCS plants and nuclear plants have a project lead-time of up to 10 years or more, with supporting investments in the supply chain stretching even further.[70]

So, nuclear build is not unique. Energy policy requires long-term thinking and planning.

---

operation in December 2015.

67. Mellen and Hollow, *No Oil*, 47.

68. Nuttall and Earp, "Nuclear Energy in the UK," 16. Also, Nuttall and Earp explained that the detailed public inquiry "provided sufficient time for the design to reach a mature stage prior to construction commencement." This minimized late design changes. Ibid., 7.

69. CCC, *Power Sector Scenarios*, 37, 46. We do not have space to analyze further the related aspect of nuclear building costs, but see the important paper: Lovering et al., "Historical Construction Costs."

70. Ibid., 108. Similarly, in 2016 IMechE, *Engineering the UK Electricity Gap*, 4, warned that with expected rising electricity demand (through population growth, more electric vehicles, etc.,): "the conclusion is that we have neither the time, resources, nor the sufficient number of skilled people to build enough CCGTs [combined cycle gas turbines] to plug this gap." The timescale is by 2025.

Objection 5: Decommissioning Costs Are Too High

People express concern about the decommissioning costs of nuclear plants that have reached the end of their life. They are convinced that costs are too high. Billions of pounds need to be spent by the taxpayer on the UK's legacy of civil nuclear power plants and other sites, e.g., the reprocessing plant at Sellafield. This is a valid concern.

First, we must remember that the Central Electricity Generating Board (CEGB) was a nationalized industry. Civil nuclear plants were built and operated by the CEGB alongside its fossil fuel power plants and hydro plants. It regularly put aside money for decommissioning its nuclear plants—this wasn't just forgotten about! However, when the industry was privatized where did this money go? We must not forget that the CEGB's Magnox reactors (all now being decommissioned) were state-owned and operated. So it seems reasonable for the state to pick up at least some of these costs.

In March 2012, a major report by Professor Gordon MacKerron, an economist and Director of SPRU (Science and Technology Policy Research) at the University of Sussex, was published on nuclear waste and decommissioning.[71] Ed Davey, then Energy Secretary in DECC under the coalition government (2010 to 2015), responding to it, announced: "New nuclear will learn from past mistakes." He noted MacKerron's remarks: "The history of managing and funding our nuclear legacy has, until very recently, been dire. Funds collected from consumers for decommissioning and waste management were diverted to other ends."[72]

"Funding future nuclear liabilities," discussed in MacKerron's report, will be different as legislation puts responsibility for funding on new operators, not UK taxpayers.[73] The responsibility for waste now lies with the Nuclear Decommissioning Authority (NDA) for these legacy reactors and other nuclear locations, e.g., Sellafield.[74] These costs are published.[75] The NDA briefly states the historical situation in which Britain developed its nuclear industry:

71. MacKerron, *Evaluation of Nuclear Decommissioning*.

72. DECC, "New Nuclear Will Learn."

73. MacKerron, *Evaluation of Nuclear Decommissioning*, 97–98.

74. The NDA was created through the Energy Act of 2008. It explains its role in "About Us": "We ensure the safe and efficient clean-up of the UK's nuclear legacy."

75. NDA, *Nuclear Provision*, sec. 3. Version updated September 1, 2016. This report noted that about two thirds of costs were met by the government with the remainder paid from revenue generated by the NDA's commercial activities. Of course, decommissioning civil nuclear plants is only part of the NDA's role.

NUCLEAR ENERGY: CONCERNS, QUESTIONS, OBJECTIONS (PART 1)

The UK nuclear industry dates from the Cold War arms race that began in the 1940s, when the focus was on producing material for nuclear weapons.

By the 1950s, its potential for peaceful uses was realised and the UK's first nuclear power stations began generating electricity for homes and businesses. During this heady era of scientific discovery, the plans for future dismantling were barely considered.[76]

Postwar Britain was a blitzed, broken, and bankrupt country—I lived in it! British historian Andrew Marr remarked: "Housing was the single post-war issue, and would remain near the top of the national agenda through the early fifties."[77] Slums desperately needed replacing. Britain owed America a fortune for its war support. Peter Hennessy's chapter "Child of the Uranium Age: The Shadow of the Bomb" describes the frightening Cold War situation. Russian H-bombs could have devastated the UK.[78] Perhaps we can appreciate somewhat how postwar Britain did too little planning for decommissioning. Hindsight is a beautiful thing.

Now the UK's new nuclear reactors will be privately owned and operated so the state and taxpayers will not pick up decommissioning bills, except indirectly through their fuel bills. Moreover, new reactors are built with future decommissioning already considered. In fact, the Generic Design Assessment of new reactors required by the Office for Nuclear Regulation and the Environment Agency/Natural Resources Wales specifically addresses decommissioning. In 2015 the ONR said it was "broadly satisfied" with Hitachi-GE's submissions on decommissioning.[79] The next stage (step 4) considers decommissioning in further detail.[80]

In March 2017, the NDA clearly laid out its detailed annual Business Plan, after a public consultation. It provides decommissioning plans until 2020, which cover its 17 nuclear sites. The NDA stated: "Our core objective is to decommission these sites safely, securely, cost-effectively and in

---

76. NDA, *Nuclear Provision*, 2. Version published February 2015.

77. Marr, *History of Modern Britain*, 73.

78. In Hennessey, *Distilling the Frenzy*, 37–78. Hennessey and I were born and lived in postwar Britain during the Cold War. The Cuban Missile Crisis (Autumn 1962) brought us so close to nuclear war.

79. ONR, *Summary Report of the Step 3*, 27. Hitachi-GE's UK Advanced Boiling Water Reactor (ABWR) is planned to be built first at Wylfa in Anglesey and then at Oldbury (north of Bristol). See my ch. 4 for a more detailed explanation of the GDA process.

80. Ibid. At the time of writing, the GDA for the UK ABWR is expected to be completed by December 2017.

a manner that protects the environment."[81] However, it admits there are some "pressing issues," including the termination of a decommissioning contract that led to legal proceedings and claims, with settlements and costs, amounting to almost £100 million. Chief Executive David Peattie recognized this was a substantial amount but considered that settling the claims prevented "costs escalating for the public purse."[82] Rightly, a government enquiry will determine what went wrong.

Conclusion

A key note to conclude is a Nuclear Industry Council quotation: "Earning and sustaining the trust and understanding of people whose livelihoods and interests are affected by your activities is an increasingly important part of corporate responsibility."[83] Public attitudes to nuclear energy are crucial and developing our understanding is vital. We need to evaluate arguments and remain alert to misinformation. Let us look at more objections.

Further Reading

Chatzis, Irene. "Decommissioning and Environmental Remediation: An Overview." *IAEA Bulletin* 57.1 (April 2016) 4–5. https://www.iaea.org/publications/magazines/bulletin/57-1.

FORATOM. *What People Really Think About Nuclear Energy.* January 2017. Brussels, Belgium: Foratom. https://www.foratom.org/publications/#topical_publications.

International Atomic Energy Agency. *Climate Change and Nuclear Power 2015.* 2015. http://www-pub.iaea.org/MTCD/Publications/PDF/CCANP2015Web-78834554.pdf.

MacKay, David J. C. *Sustainable Energy—Without the Hot Air.* Cambridge: UIT, 2009. http://www.withouthotair.com.

Nuclear Industry Association. *Decommissioning.* Briefing paper. 2016. https://www.niauk.org/wp-content/uploads/2016/09/Decommissioning.pdf.

———. *UK Advanced Boiling Water Reactor (UK ABWR) Justification Debate.* Briefing paper. April 2015. http://www.niauk.org/justification-application-uk-abwr.

———. *UK Public Opinion.* January 2016. http://www.niauk.org/images/graphics/facts_enlargements/public_op2015.pdf.

World Nuclear Association. "Supply of Uranium." December 2016. http://www.world-nuclear.org/info/Nuclear-fuel-cycle/Uranium-Resources/Supply-of-Uranium/.

---

81. NDA, *NDA Business Plan (2017 to 2020)*, 8. There is a summary in: WNN, "UK Sets Out."

82. NDA, *NDA Business Plan (2017 to 2020)*, 7.

83. NIC, *Nuclear Energy and Society*, iii.

Questions

1. What do you think about the results of the UK's Department for Business, Energy and Industrial Strategy (BEIS) public survey published in February 2017, which showed 36 percent supported nuclear energy, 20 percent opposed it, and 42 percent were neutral?

2. How do the responses for the five objections discussed above help you in assessing the future role of nuclear energy in a balanced energy mix?

3. What do you consider are suitable Christian responses in this debate and dialogue?

6

Nuclear Energy
Concerns, Questions, Objections (Part 2)

> We live at a time when emotions and feelings count more than truth, and there is a vast ignorance of science.
>
> JAMES LOVELOCK[1]

> Nuclear power is a potential safety threat, if something goes wrong. Coal-fired power is guaranteed destruction, filling the atmosphere with planet-heating carbon when it operates the way it's supposed to.
>
> BILL MCKIBBEN[2]

Introduction

WE'VE LOOKED AT SOME objections to nuclear power—let's see some more. People often raise concerns in isolation from other risks in life and options for generating electricity, without any balanced discussion. My aim is to deal fairly with objections on safety, security, safeguards, accidents, and waste.

---

1. MacKay, *Sustainable Energy*, 2.
2. Brand, *Whole Earth Discipline*, 91.

Objection 6: Safety

Are your images of nuclear safety good, bad, or dreadful? Do you think of leaking radioactive pipes, unauthorized waste discharges, and accidents? The safety of nuclear power plants is one of the public's chief concerns.[3]

To be clear, let us ask: What is nuclear safety? The IAEA definition is: "The achievement of *proper operating conditions*, prevention of *accidents* or mitigation of *accident* consequences, resulting in *protection of workers*, the public and the environment from undue *radiation* risks."[4]

At university, I learnt how to use radioactive material safely in laboratories. During my master's research, I used a large radioactive cobalt-60 source (emitting penetrating gamma rays). It was installed in a departmental basement behind substantial shielding, with a controlled-entrance safety system via a secure door and labyrinth. We were trained and authorized to use the system with its interlocks, timer, and alarm. After the all-clear to enter the inner room we took a handheld radiation monitor to check that the source was shielded and radiation levels were safe. At the same time, I wore my personal dosimeter. We knew what radiation safety entailed and followed procedures. And we had no accidents.

This training prepared me for working on a nuclear power plant. The first week (over 45 years ago) I started work, after university, one reactor had been shut down for routine maintenance, but it was about to be restarted. Before then I had taken the opportunity to join a team entering the thick prestressed concrete pressure vessel containing the boilers and reactor. Shielding separated the boilers from the reactor, but beta and gamma radiation was present. Moreover, loose radioactive material was on surfaces and in the air. We completed a full change: coveralls, boots, head covering, and respirator. We wore personal dosimeters, carried radiation monitors, and torches. After our entry into the boiler areas (not the reactor itself) we returned and followed safety procedures for leaving the area: changing, showering, and monitoring. It was an important experience for me. At university, our lecturer had explained entry procedures; now I had moved from theory to practice.

UK plant operators must comply with Acts of Parliament, regulations/rules, and the nuclear site licence conditions. The Office for Nuclear

---

3. Our focus is nuclear safety not conventional safety (e.g., slips, trips, and falls), although this is also important.

4. IAEA, *IAEA Safety Glossary* (2007 ed.), 133. Italics original.

Regulation (ONR) inspects, influences, and enforces safety and security. Discharge permissions are granted by regulators according to the plant's location.

As we saw in chapter 4, proposed new reactors are assessed through the Generic Design Assessment. Once the assessment is satisfactorily completed, operators require an approved security plan, a nuclear site licence, etc. ONR regulates the plant's construction, commissioning, operation, and maintenance. All plants have safety reviews. One particular nuclear site licence requirement is the Periodic Safety Review (PSR), conducted every ten years and anticipating the next ten years. This PSR is formally submitted to ONR for assessment and their written report. Then "ONR will either confirm with the licensee a favourable decision on the adequacy of the review, or will set out specific actions to be taken."[5] You'll recall that ONR, independent of government, holds the industry to account on behalf of the public—a significant role. The review should be useful within the organization and "should not be aimed solely or specifically at the regulator."[6] The results are summarized in an overview report.[7]

I worked on the Periodic Safety Review for one nuclear plant. Different people wrote defined sections and I authored the radioactive waste facilities section. The whole process was thorough, including good management and supervision, standards, procedures, peer reviews, meetings, site visits, and committees. Safety improvements were identified and the set of documents submitted to the regulator. Finally, ONR has many documents (too many to consider!), but periodic and interim safety reviews must follow ONR's important *Safety Assessment Principles for Nuclear Facilities*.[8]

What happens when plants age? In early 2016 EDF Energy announced lifetime extensions for four AGR nuclear sites following technical reports (see chapter 4). How was this announcement received? Journalist Robin Pagnamenta's headline read: "Ageing Nuclear Plants to Keep Lights on for

---

5. ONR, *Periodic Safety Reviews*, 16, para. 5.64. PSRs apply to decommissioning sites.

6. Ibid., 6, para. 5.3. This Nuclear Safety Technical Assessment Guide mentions the Western European Nuclear Regulators' Association (WENRA). Its objective "is to develop a common approach to nuclear safety in Europe by comparing national approaches to the application of the International Atomic Energy Agency (IAEA) safety standards" (para. 4.3).

7. Ibid., 13, para. 5.38.

8. ONR, *Safety Assessment Principles*, 13. The document is primarily aimed at ONR's inspectors to assist them in making consistent judgements (also using other documents), but it gives guidance to other people on ONR's expectations—see para. 3.

Years: Safety Fears Rise as Reactors Allowed to Operate Beyond Lifespan."[9] The people with the "safety fears" are not clearly identified. Concerns were raised by a journalist and one consultant that the ONR wouldn't manage the additional workload, but an ONR spokeswoman provided reassurance.[10] ONR reviews the technical overview reports on the plants and publishes these online.[11]

We've mentioned plants, but what about their staff? They're important. During my years in the industry I worked with men and women who were qualified, dedicated, and competent people who focused on safety. They included Christians, those of other faiths, and those with no specific religious faith. During my time in the industry I completed my first theology degree (part-time). I think people are often overlooked in debates. These people have families, contribute to the local economies, may attend churches, and are conscious of their skills and responsibilities.

People in the industry work within a nuclear safety culture. Let's see what this means to the US Nuclear Regulatory Commission (NRC): "The NRC defines nuclear safety culture as the core values and behaviors resulting from a collective commitment by leaders and individuals to emphasize safety over competing goals to ensure protection of people and the environment."[12] It expects individuals and organizations to follow a positive safety culture and identifies nine traits which are: "patterns of thinking, feeling, and behaving that emphasize safety, particularly in goal conflict situations, such as when safety goals conflict with production, schedule or cost goals."[13] To avoid confusion the NRC and nuclear industry worked together to produce the reference *Safety Culture Common Language*.[14] The Atlanta-based Institute of Nuclear Power Operations (INPO) has its own document on basic traits, used worldwide.[15]

---

9. Pagnamenta, "Ageing Nuclear Plants," 35.

10. On recruitment targets, see ONR, *Office for Nuclear Regulation Strategic Plan 2016–2020*, 26. Professor Andy Blowers was the nuclear consultant.

11. For example, see ONR, *ONR Review of NGL DNB PLEX* (August 2014) on the life extension for Dungeness B Power Station (announced earlier).

12. NRC, "Safety Culture."

13. NRC, "Safety Culture Policy Statement." NRC notes other traits may also be important.

14. Keefe et al., *Safety Culture Common Language*. See also NRC, "Safety Culture and Nuclear Reactors" and "Safety-Conscious Work Environment" (encouraging employees to raise safety concerns).

15. INPO, *Traits of a Healthy Nuclear Safety Culture*.

In the UK the safety culture is similar, and is defined as: "The assembly of characteristics and attitudes in organisations and individuals which establishes that, as an overriding priority, protection and safety issues receive the attention warranted by their significance (IAEA Safety Glossary)."[16]

Let's turn to EDF Energy, which operates fifteen UK civil reactors (providing 9.6 GWe).[17] Its document *EDF Energy Nuclear Generation: Our Journey towards ZERO HARM* demonstrates openness and transparency. The importance of safety is clearly stated:

> Nuclear safety is our overriding priority. Every one of us has a direct or indirect impact on nuclear safety and it must be at the forefront of what we do. Additionally we must ensure that radiological, environmental, industrial and fire safety are adequately controlled in support of our ambition to achieve a zero harm safety record.[18]

That's clear enough. EDF Energy's plants are inspected by UK regulators and the World Association of Nuclear Operators (WANO) visits each nuclear plant regularly for peer reviews.[19] Furthermore, EDF Energy has its own Safety and Regulation Division and each plant has a Nuclear Safety Committee giving advice and overseeing safety. So, what are their safety figures?

As regards radiation exposure of workers, the company states that nobody has received a significant uptake of radioactive material or a dose above the legal limit.[20] Actually, the legal limit is 20 mSv/year, but EDF Energy works to a company dose restriction of 10 mSv/year.[21] In 2014 the average worker's annual dose was much less than the dose received from natural background.[22]

What about public doses? EDF Energy noted:

16. ONR, *Safety Assessment Principles*, 219.

17. EDF Energy also has in the UK: coal stations, one gas station, gas storage facilities, and renewables. EDF Energy, *EDF Energy Nuclear Generation*, 5.

18. Ibid., 8.

19. WANO's mission is: "To maximise the safety and reliability of nuclear power plants worldwide by working together to assess, benchmark and improve performance through mutual support, exchange of information and emulation of best practices." WANO, *2015 Year-End Highlights*, 2.

20. EDF Energy, *EDF Energy Nuclear Generation*, 23.

21. Ibid., 37. In 2014 the highest individual dose received was 6.905 mSv.

22. Ibid., 38. In 2014 the average individual doses were 0.062 mSv (employees) and 0.112 mSv (contractors). M chapter 2 showed that the average annual dose to the UK population was 2.7 mSv, with 2.3 mSv from natural background and 0.44 mSv from artificial radiation.

> Doses to the public are a very small fraction of the legal limit and the average radiation dose due to natural background in the UK. The maximum dose received over this period (0.005 mSv in 2014) is equivalent to the natural radiation dose received during a single flight from London to Rome.[23]

This comparison to the dose received in one flight is useful. In 2012 and 2013 the maximum dose was 0.006 mSv/year while the legal limit is 1 mSv/year. In other words, public doses were a truly tiny amount. This is reassuring.

Objection 7: Security

People are concerned about terrorist threats against nuclear power plants. However, we know that security threats and attacks are not restricted to nuclear plants. The UK's Department of Energy and Climate Change (DECC) has worked with the UK's Centre for the Protection of National Infrastructure (CPNI) on "critical" infrastructure assets to reduce their vulnerability from terrorism and other security issues.[24] National infrastructure sectors, of course, include energy and civil nuclear. The World Energy Council (WEC) has drawn attention to cyber risks as a "unique concern" in energy infrastructures, including coal, oil, and nuclear plants.[25]

What do we mean by security? Well, you'll recall the UK's energy trilemma is having energy that's affordable, secure, and low carbon. Nuclear's advantage is providing secure electricity supplies. It's not intermittent, weather dependent, or season dependent (unlike solar and wind energy). However, this isn't what we mean by security here. The IAEA defined nuclear security as: "The prevention and detection of, and response to, theft, *sabotage*, unauthorized access, illegal transfer or other *malicious* acts involving *nuclear material*, other *radioactive material* or their associated facilities."[26] We'll use this definition.

Thus, the *Office for Nuclear Regulation Strategic Plan* recognized:

---

23. Ibid., 38.

24. DECC, *Overarching National Policy Statement*, 64, para. 4.15.1. See the CPNI website at http://www.cpni.gov.uk. It is worth reminding readers that DECC joined the Department for Business, Energy and Industrial Strategy (BEIS) in July 2016.

25. WEC, "Road to Resilience." Summarized in *WNN*, "Cyber Risk Must Be Managed."

26. IAEA, *IAEA Safety Glossary* (2007 ed.), 133. Italics original.

> Nuclear security and nuclear safety are complementary, but effective delivery of nuclear safety does not automatically deliver effective nuclear security. Each discipline has its own applicable UK legislation with distinct, and equally important, obligations on duty-holders.[27]

Put simply: "nuclear safety seeks to ensure risks are reduced as low as reasonably practicable," while "security aims to prevent unauthorised removal or sabotage of nuclear material/facilities and to prevent the proliferation of nuclear technologies and information."[28] Safety and security protect the public and workforce.

I hope this clarification helps. Moreover, the ONR works with other government bodies, security, and intelligence agencies. The Cabinet Office report *HMG Security Policy Framework* (2014) addressed security within HMG organizations and relevant third parties. There are overarching principles and policy priorities: information security, physical security, and personnel security (e.g., confirming identities).[29] The framework recognizes threats and the need for protection. People who say that the nuclear industry is secretive should recognize the need for security restrictions. Some documents are available online, but a balance is maintained. Terrorist attacks on New York and Washington on September 11, 2001 (9/11) raised concerns about the amount of material readily available publicly. It made terrorists' attacks easier. "It was recognised that the benefits of a culture of openness were accompanied by risks."[30]

ONR's *Finding a Balance* guides the industry on how to meet its legal requirements while protecting security. Within ONR the Civil Nuclear Security (CNS) Programme addresses security.[31] New nuclear builds are security assessed during the Generic Design Assessment. Moreover, an approved security plan is required before a nuclear site licence is granted for the first time or construction begins on a new nuclear plant.[32]

---

27. ONR, *Office for Nuclear Regulation Strategic Plan 2016–2020*, 15.

28. Ibid.

29. Cabinet Office, *HMG Security Policy Framework*, 11.

30. ONR, *Finding a Balance*, iii.

31. ONR, "Civil Nuclear Security." In July 2016, ONR reported that its new Security Assessment Principles (SyAPs) will replace the current guidance and be published in March 2017. They were published on March 31.

32. ONR, *Licensing Nuclear Installations*, 48–49. Also see ONR, *Guide to Nuclear Regulation*, 24–25.

NUCLEAR ENERGY: CONCERNS, QUESTIONS, OBJECTIONS (PART 2)

Robert F. Kennedy Jr. addresses concerns over US national security in his book *Crimes Against Nature*.[33] Kennedy doesn't start with nuclear plants, but focuses on chemical facilities and the mass loss of life/injury possible from an attack on a chlorine plant. He is concerned about federal security standards, or lack of them. Kennedy turns to the nuclear industry with equal concern directed at the government, Homeland Security, the Nuclear Regulatory Commission, and others on inadequate security.[34] It makes chilling reading.

In contrast, Gwyneth Cravens's 2007 book *Power to Save the World* looks at terrorism and effective security for nuclear plants.[35] She gives readers a guided tour discussing security with informed people. It's a chapter worth reading. Moreover, the Nuclear Regulatory Commission (NRC) states that plants "are well-protected by physical barriers, armed guards, intrusion detection systems, area surveillance systems, access controls, and access authorization requirements for employees working inside."[36] The NRC report *Protecting Our Nation* (August 2015) provides a useful discussion, including cyber security. Moreover, the US Department of Energy's National Nuclear Security Administration (NNSA) issued a comprehensive report in 2016 on global nuclear threats, including security and safeguards.[37] Also in 2016, the Washington-based Nuclear Threat Initiative issued a report by Alexandra Van Dine and colleagues that addressed nuclear power plants and warned: "Cyberattacks are a powerful tool for those who are determined to terrorize the public, undermine confidence in civilian nuclear power, or both."[38]

But let's return to the UK. Armed police are provided by the Civil Nuclear Constabulary to guard nuclear power plants.[39] It's perhaps surprising to UK citizens who aren't accustomed to seeing armed police in the streets. The Centre for the Protection of National Infrastructure (CPNI) works with the police and other bodies, e.g., the Office of Cyber Security

---

33. Kennedy, *Crimes Against Nature*, ch. 9, "National Security."

34. Ibid., 164–73.

35. Cravens, *Power to Save*, ch. 12, "Barriers."

36. NRC, "Radiation and National Security."

37. NNSA, *Prevent, Counter, and Respond*. It addresses Iran and the Joint Comprehensive Plan of Action. For a summary of the report, see *WNN*, "USA Sets Out Nuclear Security Strategy."

38. Van Dine et al., *Outpacing Cyber Threats*, 9.

39. See https://www.gov.uk/government/organisations/civil-nuclear-constabulary.

and Information Assurance, formed by the government in 2010.[40] For understanding the threats from insider attacks the CPNI has a readable personal security infographic—even employees can become threats.[41] The UK government launched its National Cyber Security Centre (NCSC) in 2016 and this has published advice for organizations on cyber security.[42]

Cyber security is now a worldwide concern for businesses. How many times have we heard companies were hacked and customers' credit card details stolen? Or national security was potentially compromised? Indeed, the World Economic Forum (WEF) in *The Global Risks Report 2016* observes that in 2014 cyber crime cost the global economy an estimated massive US$445 billion. It warns: "Businesses in all industries and of all sizes have been affected by the increased complexity, novelty and persistence of cyberattacks, with consequences ranging from the reputational to economic and legal."[43] Moreover, eight economies including the USA, list cyberattacks "as the risk of highest concern."[44] A UK government press release (May 8, 2016) warns: "Two thirds of large UK businesses hit by cyber breach or attack in past year."[45]

The nuclear industry isn't ignorant of cyber threats. Papers in the journal *Nuclear Future* address such security threats, e.g., during design of instrumentation and control systems.[46] What if malware is cleverly hidden in these systems? What damage could be done if a threat succeeds? Not necessarily death and destruction. It could aim to cause disruption, economic damage, reputation damage, blackmail, etc.[47]

The House of Commons and House of Lords Joint Committee on the National Security Strategy (HCHLJC) report *The Next National Security Strategy* (2015) was somewhat critical of the government's continuing engagement and notes: "we express our frustration at the lack of preparation

40. See the site https://www.gov.uk/government/groups/office-of-cyber-security-and-information-assurance.

41. CPNI, *Personal Security*.

42. Cabinet Office, *Prospectus*; and NCSC, "Common Cyber Attacks," which includes a case study example of an attack against the UK energy sector.

43. WEF, *Global Risks Report 2016*, 77. Estimated US$445 billion provided by the Center for Strategic and International Studies with McAfee.

44. Ibid.

45. Department for Culture, Media and Sport, "Two Thirds."

46. For example, Haines and Gaines, "Cyber Hardness"; and Silin, "Layer-Based Data Security."

47. Silin, "Layer-Based Data Security," 59.

and consultation for the next NSS."⁴⁸ It regarded the government's position as "currently too passive."⁴⁹ Nevertheless, the Cabinet Office's annual report *The UK Cyber Security Strategy 2011–2016* notes that since 2011 the government has invested £860 million in cyber security to make Britain secure. Because cyber is a top-level threat, there is an increased investment to £1.9 billion.⁵⁰ Nuclear security isn't specifically addressed, although the report states that individual and collective responsibilities are essential: government can't keep society safe unless everyone plays their role.

But let's return to EDF Energy. Their publication *Our Journey towards ZERO HARM* describes the approach to nuclear security. It makes reassuring reading, taking nuclear security seriously for personnel, physical assets, information, and the public. Policy and the management system are discussed. Security exercises are conducted and assessed. Continuous improvement is encouraged.⁵¹ On risk profile it "covers all the perceived security threats to the business from normal crime and malicious behaviour through to protestor disruption, cyber attacks and terrorism."⁵² In April 2016, EDF Energy CEO Vincent de Rivaz, speaking at the Nuclear Safety Symposium in London, identified the importance of safety and security, including new concerns such as cyber security.⁵³

Threats evolve. Caroline Baylon and colleagues produced a report at the Royal Institute of International Affairs, Chatham House.⁵⁴ They note four primary threat actors: "hacktivists; cyber criminals; states (governments and militaries); and non-state armed groups (terrorists)."⁵⁵ Major challenges identified for the nuclear industry were: industry-wide, cultural, and technical. Wide-ranging recommendations were provided for improvements. They note that after 9/11 physical security improved so:

---

48. HCHLJC, *Next National Security Strategy*, 3.
49. Ibid., 16.
50. Cabinet Office, *UK Cyber Security Strategy*, 5–6.
51. EDF Energy, *EDF Energy Nuclear Generation*, 25–26.
52. Ibid., 25.
53. WNN, "Nuclear Industry Needs."
54. Baylon et al., *Cyber Security at Civil Nuclear Facilities* (September 2015).
55. Ibid., 5. Page numbers refer to the full report; an executive summary is also available. The authors explain: "Hacktivists such as radical fringe anti-nuclear power groups might carry out a cyber attack on a nuclear facility to raise awareness of vulnerabilities. Their goal is sabotage or disruption . . ."

... the industry has now reached a high level of security (referred to as 'gates, guards and guns'). However, this very robustness may in itself make the cyber route a particularly attractive alternative for those seeking to cause damage, as it is now seen as the 'soft underbelly' of the industry.[56]

Their research findings should encourage discussion about cyber attacks and responses to them, for the benefit of our societies[57]—a commendable aim, indeed. International conferences are held on nuclear security and these have highlighted progress in security measures.[58] Finally, in December 2016 the *IAEA Bulletin* reported on nuclear security commitments and actions, including an article by May Fawaz-Huber on "How the United Kingdom seeks to enhance nuclear security with the help of IPPAS [International Physical Protection Service mission]." An IAEA team visited the UK in 2011 with a follow-up mission in February 2016. She notes: "IPPAS missions provide advice on how to improve the effectiveness of a State's physical protection regime, either nationally or at facility level."[59] Let us turn to safeguards.

Objection 8: Safeguards

We've considered safety and security; next it's nuclear safeguards. Actually, these three (the three Ss) go together on protecting the public and environment from exposure to ionizing radiation and harmful effects. Safeguarding is the activity designed to prevent proliferation of nuclear weapons. These three essential areas for nuclear sites should be integrated, as Williams's 2013 *Nuclear Future* article explains.[60]

---

56. Ibid., 14.

57. Ibid., 37. For recommendations, see Baylon et al., *Cyber Security at Civil Nuclear Facilities*.

58. IAEA, "International Conference on Nuclear Security," mentioned the successful first International Conference on Nuclear Security (ICNS), "Enhancing Global Efforts," held in Vienna in July 2013. The second ICNS, "Commitments and Actions," was held in Vienna on December 5–9, 2016.

59. Fawaz-Huber, "How the United Kingdom Seeks to Enhance Nuclear Security," 12.

60. Williams, "Nuclear Safety, Security" (2013), 33. When he wrote this article Laurence Williams was Professor of Nuclear Safety and Regulation in the University of Central Lancashire, UK. For a more recent 2017 paper on integrating the 3S (or Triple S), see Rodger et al., "Nuclear 3S." The five authors work at the UK's National Nuclear Laboratory.

Readers may recall the IAEA's involvement in negotiations with Iran over concerns that it was secretly developing nuclear weapons. The IAEA has a global responsibility for safeguards. Put simply: "Nuclear safeguards are measures to verify that countries comply with their international obligations not to use nuclear materials for nuclear explosives."[61] This includes materials such as uranium enriched with uranium-235, and plutonium-239. There is an international Treaty on the Non-Proliferation of Nuclear Weapons which many states have signed agreeing that IAEA can apply safeguards on their nuclear material. Verification of safeguards includes visits by international inspectors. However, the Office for Nuclear Regulation website notices that after the 1990s Gulf War, although Iraq had signed the treaty, it was found to have "managed to pursue a substantial nuclear weapons programme."[62] Consequently, the IAEA safeguards system was strengthened.

Besides the IAEA, the European Union has safeguards provisions applied by the European Commission on member states (called Euratom safeguards). Without going into fine detail, the UK provides information to the IAEA via the European Commission, including information on its civil nuclear plants.[63] The IAEA and European Commission can inspect them if they wish. And ONR works closely with these teams and the UK government.[64] Specifically, the ONR publishes annual civil stocks of plutonium and uranium, including amounts of plutonium in spent fuel and highly enriched uranium (HEU).[65] The IAEA provides extensive information.[66] Moreover, on current trends the Agency notes: "The global nuclear landscape is changing rapidly and will likely continue to do so. Every day—across the world—more nuclear facilities and material come under IAEA safeguards. Nuclear power is expanding—in countries already using it, as well as in States introducing it."[67]

---

61. ONR, "What Are Nuclear Safeguards?" See also ONR, "Basic Safeguards Glossary."

62. ONR, "What Are Nuclear Safeguards?"

63. See ONR, "IAEA Safeguards in the UK."

64. ONR, "Guide to Nuclear Regulation," 31.

65. ONR, "Annual Civil Plutonium and Uranium Figures." HEU is defined as uranium enriched to 20 percent or more in uranium-235.

66. IAEA, *IAEA Safeguards*, 6. IAEA's Department of Safeguards had around 850 people. During 2014 inspections exceeded 2,700 visits, samples were collected, and satellite images analyzed.

67. Ibid., 18.

EDF Energy's *Our Journey* doesn't specifically address nuclear safeguards, although it is involved. Nuclear safeguarding is a global issue. Much has been done and much remains to be done, as a 2016 White House press briefing summarizes.[68]

What then should we conclude on new installations in the UK? Will the fear of proliferation—through material being diverted to produce nuclear weapons—stop the UK from choosing to build new plants? Civil nuclear fuel uses low-enriched uranium, which cannot make a nuclear weapon. Any plutonium in spent fuel would need to be first obtained by stealing the highly radioactive fuel and using complex chemical separation facilities to obtain plutonium before fabricating a bomb. What does the World Nuclear Association (WNA) say? It notes:

> Civil nuclear power has not been the cause of or route to nuclear weapons in any country that has nuclear weapons, and no uranium traded for electricity production has ever been diverted for military use. All nuclear weapons programmes have either preceded or risen independently of civil nuclear power*, as shown most recently by North Korea. No country is without plenty of uranium in the small quantities needed for a few weapons.[69]

The ONR reported that in 2014 there were almost 220 inspections performed by Euratom and IAEA safeguards inspectorates. They concluded there was "no diversion of safeguarded material in the UK."[70]

Objection 9: Accidents

Another objection to nuclear energy is accidents. Of course, accidents aren't restricted to modern times. In ancient Rome, the satirist Juvenal (c. AD 55–140) mentions the danger of fires and collapsing houses.[71] Jesus mentions a tower collapsing: "Or those eighteen who died when the tower in Siloam fell on them—do you think they were more guilty than all the others living in Jerusalem?"[72]

---

68. White House, "Nuclear Security Summits."

69. WNA, "Safeguards to Prevent Nuclear Proliferation," Conclusion. *An exception may be South Africa.

70. ONR, *Office for Nuclear Regulation Annual Report and Accounts 2014/15*, 44.

71. Juvenal, *Sixteen Satires*, Satire III, lines 190–231 (pp. 93–95).

72. Luke 13:4.

NUCLEAR ENERGY: CONCERNS, QUESTIONS, OBJECTIONS (PART 2)

In chapter 3 I described the major accident at Aberfan, South Wales, on October 21, 1966. A coal tip—waste created from local coal mines—crushed the village's junior school, some houses, and a farm, killing 116 children and 28 adults. Then there was the industrial accident in Bhopal, India, during the night of December 2, 1984, when the Union Carbide insecticide plant released tons of methyl isocyanate gas. The BBC noted: "Half a million people are exposed to the gas. Within three days, 8,000 are dead. Thousands more die in the months afterwards."[73] Survivors suffered—compensation claims continue. In April 2016, an accident at Paravur town (Kerala State, India) happened in a fireworks display competition during a Hindu festival at a temple complex. Explosions and fires tragically killed at least 106 people with hundreds injured, stated *The Times* newspaper.[74]

But two energy-related accidents spring to mind. First, the Piper Alpha accident on the UK's North Sea oil platform (July 6, 1988) killed 167 workers in the gas leak, explosion, and fire. I knew someone whose relative died. Second, remember the Gulf of Mexico disaster on BP's *Deepwater Horizon* oil rig? This began on April 20, 2010, killing 11 men in the explosion and fire and releasing massive amounts of oil and gas. The sea and shore were polluted, affecting people's lives and the environment. Bob Cavnar analyzes the accident. He concludes we must move away from fossil fuels but also "toward alternative sources of energy, *including* nuclear power, if we can make it safer..."[75] I could go on.

But uniquely associated with nuclear plants are fears of radiation exposure and cancer. Radiation accidents are unique to them, aren't they? Well, no! Just because we hear about Three Mile Island (US, 1979), Chernobyl (Ukraine, 1986) and Fukushima (Japan, 2011) this may distort our perception. Jean-Claude Nénot remarks:

> Radiation accidents may be classified into two categories: 'radiological accidents', involving installations that use the properties of radiation, and 'nuclear accidents', resulting from the direct use of atomic fission and affecting nuclear facilities and weapons. Both may expose the workforce and/or members of the population.[76]

73. A. Francis, "Why Are Bhopal Survivors?"

74. Tomlinson, "'Banned' Firework Competition," 32.

75. Cavnar, *Disaster on the Horizon*, 181. Italics original. See ch. 12, "The Aftermath: What Do We Do Now?" For further information, see Deepwater Horizon Study Group, *Final Report*.

76. Nénot, "Radiation Accidents," 301 n. 2. At the time of writing his article Jean-Claude Nénot declared his status as the Institut de Radioprotection et de Sûreté Nucléaire (2008)

Nénot rightly observes that industrial and medical uses of radiation have expanded significantly since the Second World War. Accidental overexposures have caused deaths and injuries of the public and workforces. He classifies severe radiation accidents over the last 60 years (where there is immediate or delayed recognition of the event and secret incidents, military or political). He assesses: "(1) at least 600 events caused significant radiation exposure of about 6000 individuals, (2) about 70 serious accidents resulted in one or more deaths each, and (3) a total of 200 lethal issues were due to ARS [acute radiation syndrome]."[77] Accidents occurred in medical situations (e.g., administering the incorrect amount of gold-198 as a radiopharmaceutical proved lethal), loading a radiotherapy source, entering rooms when a large radioactive source was exposed, finding a lost industrial radiography source, and keeping it close enough to irradiate people.[78]

For example, the Goiânia (Brazil, 1987) radiological accident concerned a derelict radiotherapy clinic with an abandoned teletherapy machine containing a radioactive source (cesium-137). This source was removed (by people ignorant of the danger), ruptured, and the remnants were sold as scrap. As the source glowed in the dark people were fascinated and fragments of the source were distributed, thus spreading radioactive contamination to people, property, and land.[79] Four people died and 28 had radiation burns. Decontamination produced a considerable amount of radioactive waste from soil and demolished houses. IAEA's 1988 report stated it was: "one of the most serious radiological accidents" to date.[80] Once alerted, the authorities acted swiftly. International cooperation followed. Lessons were learnt, not least on effective communication:

> There were two distinct phases in the reaction of the communications media (the press, radio and television). The first was characterized by sensationalism, misinformation and criticism of the authorities. In the second phase there was a much more responsible coverage of events, seeking to inform the public and describing more clearly what was happening...[81]

---

and the International Commission on Radiological Protection Main Commission (2001).

77. Ibid., 303.
78. Ibid., 304–5.
79. IAEA, *Radiological Accident in Goiânia*, 1.
80. Ibid., foreword.
81. Ibid., 116.

NUCLEAR ENERGY: CONCERNS, QUESTIONS, OBJECTIONS (PART 2)

Let's turn to nuclear accidents. On October 10–11, 1957, a fire occurred in the UK's Windscale Reactor No. 1 pile (a military reactor for plutonium production). Widespread radioactive contamination spread from the plant into the atmosphere and was deposited on the ground. Cows' milk was banned for some weeks from farms in an area of about 200 square miles (518 km^2) because of radioactive iodine-131. At the time, I was a child living with my family in the city of Liverpool more than 30 miles (48.5 km) from Windscale. Radioactive iodine concentrates in the thyroid, so the drinking water in our two reservoirs was tested for iodine-131. Activity levels were found to be low.[82]

An editorial article in the *Journal of Radiological Protection* (2007) looking at the accident 50 years later estimated about 100 fatal cancers and about 90 non-fatal cancers in total from the accident.[83] However, a 50-year follow-up of the workers involved in extinguishing the Windscale fire and clean-up operations concluded that the analysis "does not reveal any measurable effect of the fire upon their health," and "does provide reassurance that no significant health effects are associated with the 1957 Windscale fire even after 50 years of follow-up."[84] It wasn't a trivial accident but was rated a Level 5 accident within the International Nuclear and Radiological Event Scale (INES).[85] More iodine-131 was released than at Three Mile Island.

At Three Mile Island (1979) a loss of coolant caused the fuel to overheat and melt (a core meltdown). Many people self-evacuated from the area. President Jimmy Carter established the Kemeny Commission. Its overview stated, concerning the accident's severity: ". . . we conclude that in spite of serious damage to the plant, most of the radiation was contained and the actual release will have a negligible effect on the physical health of individuals. The major health effect of the accident was found to be mental stress."[86] Lessons learnt provided many recommendations.

---

82. Dunster et al., "District Surveys," 224. Lovelock, *Revenge of Gaia*, 100, describes his experience as a scientist finding out about the Windscale fire.

83. Wakeford, "Windscale Reactor Accident," 213–14. Radiation exposure from iodine-131 and polonium-210.

84. McGeoghegan et al., "Mortality and Cancer," 430. The authors were from the Epidemiology Group, Westlakes Research Institute, Cumbria, UK. For more recent, 2016 studies, see Jones, "Health Effects of the Windscale Pile fire," and McNally et al., "A Geographical Study of Thyroid Cancer Incidence."

85. For an explanation of the scale, see IAEA, *INES*.

86. Kemeny Commission, *Report of the President's Commission*, 12. Published in October 1979. The chairman, John G. Kemeny, was President of Dartmouth College.

At the Chernobyl nuclear power plant (Ukraine) on April 26, 1986, the Unit 4 reactor experienced explosions in its core, which destroyed the core and buildings. Radioactive material was ejected from the site and the core caught fire. Winds spread contamination over Europe in the world's most severe reactor accident.[87] Mikhail Balonov stated: "The explosions were caused by gross breaches of the operating procedures by staff and technical inadequacies in the safety systems."[88] The unstable reactor design would not have received a nuclear site licence in the UK. Radioactive iodine-131 was absorbed in fresh milk, and people consuming this received high thyroid doses. Here Balonov noted: "In the first few weeks, management of animal fodder and milk production (including prohibiting the consumption of fresh milk) would have helped significantly to reduce the doses to the thyroid due to radioiodine."[89] Early countermeasures were flawed. In another article, on protecting "the inexperienced reader," Balonov agreed that the accident produced many harmful effects (on people and the environment) but he robustly rejected misleading exaggerated claims of deaths based upon a mistaken methodology. Such mistakes ("myths"), he argued, could lead to erroneous responses to the Fukushima nuclear plant accident.[90]

Although the focus is often on radiation, Evelyn Bromet notes that the Chernobyl Forum (2006) concluded: "mental health problems constituted the largest public health problem."[91] And she recognized its importance in her 25th-anniversary review. Mental health issues impacted on physical health. Bromet recommended that mental and physical aspects be included in health programs for Fukushima populations.

The accident at Fukushima (Japan) on March 11, 2011, was caused by a large underwater earthquake. The operating reactors shut down automatically, but the massive tsunami washed over the protective wall and flooded the site about one hour later. Electrical power was eventually lost, reactors overheated, and hydrogen explosions occurred. Significant quantities of

---

87. Balonov, "Chernobyl Accident," 27–28. Balonov (Institute of Radiation Hygiene, St, Petersburg, Russia) was a consultant to the United Nations Scientific Community on the Effects of Atomic Radiation (UNSCEAR) for Chernobyl.

88. Ibid., 28.

89. Ibid., 30.

90. Balonov, "On Protecting the Inexperienced Reader," 181. Specifically, he addresses Yablokov et al., with an overestimation of 800,000 deaths in 1987–2004.

91. Bromet, "Mental Health Consequences," N74. Evelyn Bromet was at the Department of Psychiatry and Behavioral Science, Stony Brook University School of Medicine, Stony Brook, New York (US).

radioactive material were released (although much less than at Chernobyl). Countermeasures were introduced (including evacuations). The UK Secretary of State for Climate Change called for an investigation on the implications for the UK industry. Although this report saw "no fundamental safety weakness" in the UK's industry, many recommendations suggested safety improvements.[92] The nuclear industry addressed them. Doses among the public outside the exclusion zone were generally low. In fact, high school students and teachers in Fukushima, France, Poland, and Belarus participated in an investigation comparing annual external individual doses in Japan with European regions. This demonstrated: "participants in Fukushima and Belarus are well within the terrestrial background radiation levels of other regions/countries."[93]

Shortly after Fukushima, journalist George Monbiot wrote: "You will not be surprised to hear that the events in Japan have changed my view of nuclear power. You will be surprised to hear how they have changed it. As a result of the disaster at Fukushima, I am no longer nuclear-neutral. I now support the technology."[94]

What about EDF Energy's UK record? In *Our Journey towards ZERO HARM* the company reports Nuclear Safety Events using the International Nuclear and Radiological Event Scale (INES) between Level 1 ("an anomaly with no impact on the safety of the general public or workforce") and Level 7 (major accident). The number of Level 1 events were: 15 in 2012, 10 in 2013, and 5 in 2014. In the last 5 years no events higher than Level 1 have occurred.[95] So no accidents occurred. The events were reported by EDF Energy and have been published by both the company and the Office for Nuclear Regulation.[96]

Recently, the Nuclear Fuel Cycle Royal Commission (NFCRC) of the government of South Australia showed a positive attitude towards nuclear energy:

---

92. ONR, "Fukushima Lessons Learned."

93. Adachi et al., "Measurement and Comparison," 65. The total number of students and teachers in the 2014 study was 216.

94. Monbiot, "Going Critical."

95. EDF Energy, *EDF Energy Nuclear Generation*, 36. See IAEA, *INES*, for more information.

96. See EDF Energy, "Office for Nuclear Regulation Publishes" and "Reporting of Safety Events." Also, see the report ONR, *Events Reported*, which puts the information in the public domain.

LET THERE BE LIGHT!

The Commission looked closely at reactor safety and the major accidents associated with nuclear power plants. While acknowledging the severe consequences of such accidents, the Commission has found sufficient evidence of safe operation and improvements such that nuclear power should not be discounted as an energy option on the basis of safety.[97]

Finally, while considering occasional (though certainly tragic) nuclear accidents, remember also the continuing destructive consequences of routinely burning fossil fuels. Climate change causes sea-level rising and irretrievable loss of land around coasts and estuaries. Small islands and agricultural land will disappear, and coastal cities will be threatened. Christian climate scientist Sir John Houghton shows that millions of people will be displaced.[98]

Objection 10: Waste

Stewart Brand, a pro-nuclear environmentalist, notes: "A common refrain against nuclear goes, 'Solving the problem of waste storage is so difficult, not a single geological repository for nuclear waste is operating anywhere in the world.'"[99] He replies: "But there is one in the United States, in operation burying radioactive stuff since 1999. WIPP, the Waste Isolation Pilot Plant, in New Mexico..."[100] This is a geological disposal facility for US military waste, which must be distinguished from civil nuclear waste and the ongoing Yucca Mountain (Nevada) repository disputes.[101] The US Department of Energy's waste repository takes radioactive material from military defense and buries it deep underground within the salt bed. It is a permanent structure for long-lived radioactive waste that is either "contact-handled" (low-activity waste)

---

97. NFCRC, *Nuclear Fuel Cycle Royal Commission Report*, xiv. Its Preface (xi) states: "The Nuclear Fuel Cycle Royal Commission was established by the South Australian Government on 19 March 2015 to undertake an independent and comprehensive investigation into the potential for increasing South Australia's participation in the nuclear fuel cycle..."

98. Houghton, "Sustainable Climate," 15. A 1 meter rise in sea level will displace about 10 million people in Bangladesh alone. For an analysis of climate change causing earthquakes, tsunamis, and volcanoes, see McGuire, *Waking the Giant*. When he published his book Bill McGuire was Professor of Geophysical and Climate Hazards at University College London.

99. Brand, *Whole Earth Discipline*, 105.

100. Ibid.

101. On disputes, see http://www.yuccamountain.org. WIPP is discussed in Cravens, *Power to Save*, ch. 18, "The Gigantic Crystal."

or "remote-handled" (higher activity waste). Its website states that 22 generator sites nationwide have been cleaned up.[102] Although WIPP had been operating since 1999 it was temporality closed down after two unrelated events in February 2014. Two Accident Investigation Boards issued reports.[103] After successful tests and actions towards restart operations, the US Department of Energy (DOE) authorized the operator to resume waste emplacements in late December 2016. In January 2017, the first waste emplacement occurred from waste already stored onsite at WIPP. At the time of writing, shipments to WIPP from generator sites were expected to commence in April 2017, according to the DOE, and this occurred as planned.[104] My point here? A US deep geological repository operating for many years takes military radioactive waste for permanent burial. This is an example of what has been achieved as countries progress in waste disposal options for civil facilities.

Discussions on national, regional, and international repositories recognize that some countries could share facilities, with multiple benefits.[105] European Council Directive 2011/70/EURATOM states requirements for disposal of radioactive wastes—all countries are responsible for their waste. The ERDO Working Group comments: "The amounts of radioactive wastes from nuclear power and industrial, medical and research activities are comparatively small, but no country in the world has yet implemented a comprehensive disposal solution for all of its national waste arisings."[106] Yet, the group recognizes advanced programs in Finland, France, and Sweden. More recently, in November 2016 permission was given by the Finnish Radiation and Nuclear Safety Authority (Stuk) for the waste company Posiva to construct a repository for spent fuel in Finland. *World Nuclear News* reported: "The excavation of an underground used nuclear fuel final disposal facility at Olkiluoto, Finland, is set to begin next month."[107] We will need to see what happens here. In November 2016 Russia began disposal of low-level and intermediate nuclear waste in its near-surface final waste

102. See WIPP, "About WIPP."

103. WIPP, "WIPP Recovery Plan."

104. *WNN*, "Waste Shipments to WIPP Expected to Resume Soon." See WIPP, "WIPP Update: April 10, 2017," and *WNN*, "WIPP Waste Shipments Resume."

105. For example, on shared solutions there is the European Repository Development Organisation (ERDO). See http://www.erdo-wg.eu/Home.html. For issues/ethics on shared repository facilities, see Taebi, *Good Governance of Risky Technology*.

106. ERDO Working Group, *Shared Solutions*, 1.

107. *WNN*, "Construction to Start on Finnish Repository." Also, see Posiva, "First Excavation Works."

repository, the first one to become operational.[108] The UK completed public consultations over the siting of a national geological disposal facility (I attended one) and published its results in April 2016.[109]

May 2016 brought an unexpected change when the Nuclear Fuel Cycle Royal Commission in Australia published its positive report with 12 recommendations on nuclear energy. They included considering establishing storage and disposal facilities for used nuclear fuel and intermediate-level radioactive waste.[110] Other countries could use the facilities. The WNA remarked that the report "has fundamentally changed the nature of the global nuclear waste discourse."[111] Indeed it has. However, in November 2016 a Citizen's Jury discussion resulted in the decision that two thirds of the 300 jurors rejected the proposal while the remaining third opted for more dialogue. Reasons given for the negative reaction include inquiring whether the question they were asked was poorly worded and if the time to discuss the issue was too short.[112] We'll see how this develops.

People express concern about radioactive waste from nuclear plants. However, it's helpful to put concerns in context. All waste concerns us—not just nuclear. For example, asbestos, carcinogens, and heavy metals that are buried won't undergo radioactive decay—they'll remain there forever. Let us be clear: civil plants are not the only producers of radioactive waste. For example, as discussed in chapter 2, some hospital discharges are radioactive and industries working with naturally occurring radioactive material (NORM) generate wastes.[113]

The question "Isn't the waste from nuclear reactors a huge problem?" was answered by David MacKay. First, he noted that in the UK the volumes are relatively small compared with the waste from coal-fueled power

---

108. *WNN*, "Near-Surface Final Waste Repository."

109. RWM, "Consultation Outcome." RWM is a subsidiary of the UK's Nuclear Decommissioning Authority.

110. NFCRC, *Nuclear Fuel Cycle Royal Commission Report*, 169.

111. WNA, "Royal Commission's Conclusions."

112. See the discussion by the solicitor Edward de la Billiere, "Going Nuclear." The question was: "Under what circumstances, if any, could South Australia pursue the opportunity to store and dispose of nuclear waste from other countries?" They had six days to examine the evidence.

113. For more information, see DECC et al., *Strategy for the Management of Solid Low Level Radioactive Waste*. See Simms, "Naturally Occurring Radioactive Material," for volumes of NORM waste. Also see DECC et al., *Strategy for the Management of Naturally Occurring Radioactive Material (NORM) Waste*.

NUCLEAR ENERGY: CONCERNS, QUESTIONS, OBJECTIONS (PART 2)

stations. Most waste is low-level, 7 percent is intermediate-level, and just 3 percent is high-level waste. He calculated amounts per capita and compared them to a wine bottle. Low-level waste is 760 ml; intermediate waste is 60 ml; and high-level is just 25 ml. Now, he noted, this total amount of waste per person per year is just over one wine bottle. And high-level waste is just 25 ml! In comparison, we see the vast difference per year per person of municipal waste (517 kg) and hazardous waste (83 kg).[114] Even the future amounts, he showed, are relatively small.

Nevertheless, "Where's it going to go?" Well, we shouldn't do what coal-fueled power stations and gas plants do—discharge CO_2 into the atmosphere (multiple millions of tons annually, for decades upon decades). Such practices came back to bite us with climate change—increasing levels in our atmosphere and acidified oceans. Also, other chemicals are discharged (e.g., mercury, arsenic, sulfur, and nitrous oxides). Moreover, discharges and coal ash are radioactive (but nobody seems to bother about these).[115] Fossil fuel plants in normal operation continue discharging uncontrolled waste which damages our planet! Nuclear plants store waste for safe disposal. In the UK, low-level waste is buried. Intermediate-level and high-level waste will be buried in a future underground repository.

Even if people object to companies building new nuclear plants, the question remains as to how the legacy waste (civil and military) produced so far can be stored safely. It can't be ignored—it needs storing somewhere. One option being considered is deep boreholes.[116] However, two uses for radioactive waste have been suggested recently. First, in India, scientists have found a way to use a solid form of radioactive cesium-137 (with a half-life of 30.2 years) instead of cobalt-60 (half-life of 5.3 years) to irradiate donated blood used in transfusions and so prevent disease transmission. The benefit is the longer half-life: it means that the radioactive sources do not need to be replaced as frequently. The intention is to use cesium-137 instead of cobalt-60

---

114. MacKay, *Sustainable Energy*, 169–70. See fig. 24.13.

115. Cravens, *Power to Save*, 196–98, discusses releases from burning coal. Kennedy, *Crimes Against Nature*, 126–28, discusses coal-burning plants, mercury discharges, and adverse effects on public health. For a recent assessment of radiation exposures from electricity generation, including coal plants, see the February 2017 summary: United Nations Information Service, "New UN Study Assesses Radiation Exposure"; and *WNN*, "UNSCEAR Studies Radiation."

116. Travis, "Deep Borehole Disposal" (2016). Karl Travis is a reader in computational physics and nuclear waste disposal at the University of Sheffield, UK. Also discussed by Cartlidge, "US Set for Borehole Field Trial."

in other medical applications, e.g., for sterilizing equipment.[117] Second, University of Bristol scientists have constructed a prototype battery that uses radioactive carbon-14 (half-life 5,730 years) to run what is called a "diamond" nuclear-powered battery.[118] Instead of carbon-14 being radioactive waste, it could end up in a clean electricity battery that lasts a long time. We'll need to see how these are developed and whether they became viable.

Finally, although space does not permit detailed discussion, the treatment of spent nuclear fuel and high-level waste in the US is an ongoing important issue requiring urgent attention. Indeed, in March 2017 the United States Nuclear Infrastructure Council (USNIC) observed: "Today, the Nation's nuclear waste management program stands at an impasse, largely due to universally recognized political reasons." Besides this impasse costing US taxpayers dearly, "the continued stalemate is damaging America's international standing on issues of nuclear safety, nonproliferation and security."[119] USNIC provides recommendations and calls for urgent action by Congress, the Trump administration, and the US DOE. Meanwhile, *World Nuclear News* reported on March 31, 2017 about US waste issues. First, the company Holtec International had submitted a license application to the US NRC to store used nuclear fuel at a below-ground storage site in New Mexico, thus enabling nuclear plant operators to dispose of their fuel offsite, pending a permanent disposal facility. Second, the Trump administration was talking to stakeholders on restarting the Yucca Mountain disposal project and requested finances for the licensing process. Energy Secretary Rick Perry visited its Nevada site in March.[120]

Conclusion

We've examined objections on safety, security, safeguards, accidents, and waste. To be sure, we live in a world where these things do concern us, but

---

117. *WNN*, "India Creates Medical Supplies."

118. *WNN*, "British University Unveils."

119. USNIC, *USNIC Backend Working Group*, 1. USNIC is a business consortium that supports new nuclear energy.

120. *WNN*, "New Mexico Used Fuel Project." Also, see the report in NEI, "New Mexico Interim Storage Facility Plan." For a useful discussion, see Peterson, "Spent Nuclear Fuel is Not the Problem." Professor Per F. Peterson is in the Department of Nuclear Engineering, University of California Berkeley, Berkeley, CA. He asks (413) "And how might scaring people about the wrong problem (spent fuel) result in poor decisions and policy?" The NEI urged the US congress to address nuclear waste, see *WNN*, "US Administration Urged."

by acting wisely, working together, and loving our neighbors we can make the world better. If we want electricity—and we do—then all electricity generation options have pros and cons. We need to prevent catastrophic climate change, and nuclear energy is a powerful ally in moving towards a decarbonized world.

Further Reading

EDF Energy. *EDF Energy Nuclear Generation: Our Journey towards ZERO HARM. Summary of Our Nuclear Safety and Waste Policies and Management Systems.* September 2015. https://www.edfenergy.com/sites/default/files/our_journey_towards_zero_harm_2015.pdf.

———. "Virtual Tours: Nuclear Generation and Safety." Video. https://www.edfenergy.com/virtual-tours/nuclear-safety. (There is a selection of virtual tours, e.g., "How we generate nuclear electricity.")

Global Nexus Initiative. *Evolving Nuclear Governance for a New Era: Policy Memo and Recommendations.* April 2017. Report based on discussions of the Global Nexus Initiative (GNI) Working Group at its June 2016 workshop in Washington, D.C. http://globalnexusinitiative.org/wp-content/uploads/2017/04/GNI-Policy-Memo-3.pdf. (The Global Nexus Initiative uses experts "to examine the complex challenges posed by the intersection of climate change, energy demand and global security," p. 1)

International Atomic Energy Agency. *IAEA Safety Glossary: Terminology Used in Nuclear Safety and Radiation Protection.* 2007 ed. http://www-pub.iaea.org/MTCD/publications/PDF/Pub1290_web.pdf.

———. *INES: The International Nuclear and Radiological Event Scale.* 08-26941/E. n.d. https://www.iaea.org/sites/default/files/ines.pdf.

———. *Linking Nuclear Power and Environment: Safe, Secure, Sustainable Nuclear Power.* 2013. https://www.iaea.org/sites/default/files/np0613.pdf.

Lucas, Edward. *Cyberphobia: Identity, Trust, Security and the Internet.* London: Bloomsbury, 2015. http://www.edwardlucas.com/.

Nuclear Industry Association. "Facts and Information for Nuclear Energy." 2016. http://www.niauk.org/facts-and-information-for-nuclear-energy.

———. "Waste Management." 2016. https://www.niauk.org/industry-issues/waste-management/.

Office for Nuclear Regulation. "Civil Nuclear Security." September 21, 2016. http://www.onr.org.uk/ocns/.

———. "Fukushima Lessons Learned: UK Action Plan Published." December 31, 2012. http://news.onr.org.uk/2012/12/fukushima-lessons-learned-uk-action-plan-published/.

Questions

1. Identify the five programs that the World Association of Nuclear Operators (WANO) provides to maximize the safety and reliability of nuclear plant. Note down what you consider to be important aspects of WANO's role in each program. Use their *2015 Year-End Highlights Report* (http://www.wano.info/en-gb/library/highlightsreport).

2. Use the Further Reading list above to examine one or two objections covered here.

3. Watch William Nuttall's lecture "Britain, Nuclear Energy and the Future," the IET Clerk Maxwell Lecture 2015 at the Royal Institution, London, March 15, 2015 (https://www.youtube.com/watch?v=wCviKi5aD-I). In what ways does this help you?

7

Grasping Our Future!

We need to promote authentically Christian responses and practical action in the arena of public policies, to ensure that science and technology are used not to manipulate, distort and destroy, but to preserve and better fulfil our humanness, as those whom God has created in his own image.

LAUSANNE MOVEMENT[1]

Introduction

IN CHAPTER 1 WE saw that God created a universe that is nuclear powered and naturally radioactive. Then chapter 2 introduced natural and artificial radioactivity. This is the context in which, I suggest, Christians consider nuclear energy. For us, God is creator and sustainer. Colossians 1:15–18 informs us about Jesus "all things have been created through him and for him," and "in him all things hold together."[2] Christians throughout the generations are followers of this risen Jesus. When Jesus walked on Earth he taught us to love God and our neighbors as ourselves. Further, the Lord's prayer teaches us to

---

1. Lausanne Movement, *Cape Town Commitment* (2011), §6. The Lausanne Movement is an international evangelical Christian movement founded by the American evangelist Dr. Billy Graham and the British evangelical leader Dr. John Stott. Its first congress in 1974 exceeded 2,400 participants from 150 nations. They met in Lausanne, Switzerland. See https://www.lausanne.org/about-the-movement.

2. Bauckham, *Bible and Ecology*, ch. 5, "From Alpha to Omega" has a reflection on these verses and other New Testament verses.

pray to our heavenly Father that his kingdom come and his will be done upon Earth, as it is in heaven. The Earth is important to God and human beings are part of the "community of creation," as New Testament scholar Richard Bauckham states: "We need to realise more fully the biblical sense in which humans are fellow-creatures with other creatures."[3] When we say the Lord's prayer about God's kingdom coming on Earth we are part of the unfolding story or biblical narrative. Therefore, this chapter summarizes the role and benefits of nuclear energy in our future on Earth. Will we be supporters?

Christians, Climate Change, and Paris COP21

The agreement reached in December 2015 at COP21 in Paris came as a welcomed surprise. Unexpectedly, the final agreement was stronger than originally anticipated. The original focus was on limiting future temperature rises to 2.0 °C, but delegates decided upon:

> Holding the increase in the global average temperature to well below 2 °C above pre-industrial levels and to pursue efforts to limit the temperature increase to 1.5 °C above pre-industrial levels, recognizing that this would significantly reduce the risks and impacts of climate change.[4]

The Paris Agreement brought international positive responses and rejoicing. The Christian community played an important role in encouraging this agreement and demonstrated the positive influence of the global church.[5] Climate conversations are no longer focused merely on economics/politics, but address morality and ethics.[6] However, much remains to be done in, for example, mitigation (reducing emissions) and adaptation (addressing the problems caused by climate change).[7] Christian involvement

---

3. Ibid., preface.

4. UNFCCC, *Adoption of the Paris Agreement*, Article 2.1(a). COP21 is the 21st session of the UNFCCC Conference of the Parties.

5. This is the view of E. Brown, *Climate Change after Paris*. Ed Brown gave the Lausanne Global Analysis. He is the Lausanne Catalyst for Creation Care, and Director of the Lausanne/WEA Global Campaign for Creation Care and the Gospel.

6. Also, Ipsos MORI, *Public Attitudes to Science 2014*, 169, noted: "people want scientists to explain their intentions more, and want to know that scientists consider the social and ethical implications of their work."

7. E. Brown, *Climate Change after Paris*, 12–13.

is what Tom Wright argues for in his book *God in Public*: Christians should show their faith in the public square—their faith is not just private.[8]

The Paris Agreement sensibly invited the Intergovernmental Panel on Climate Change (IPCC) to submit by 2018 a special report on the figure of 1.5 °C with associated global warming impacts and pathways of greenhouse emissions.[9] However, in 2016 European academics produced an analysis of key impacts (e.g., on extreme weather, sea level, and coral reefs) due to global warming of the difference between 1.5 °C and 2.0 °C. They stated:

> Our results reveal substantial differences in impacts between a 1.5 °C and 2.0 °C warming that are highly relevant for the assessment of dangerous anthropogenic [i.e., human] interference with the climate system. For heat-related extremes, the additional 0.5 °C increase in global-mean temperature marks the difference between events at the upper limit of present-day natural variability and a new climate regime, particularly in tropical regions.[10]

So, every effort should be made to limit temperature rise to 1.5 °C. Nuclear energy should be accepted within a tool box of technologies. What some attendees may not have appreciated at COP21 in Paris was that the electricity used there was clean, and low carbon, from France's large nuclear reactor fleet (around 80 percent of France's electricity is currently nuclear generated).[11]

Now the good news is that COP21 came into force in November 2016, but *World Nuclear News* reported in December 2016 that László Varró, the International Energy Agency's chief economist, had many concerns. Summarizing Varró, *WNN* stated: "Wind and solar power are transforming the electricity industry, but not fast enough to put the world on track for the UNFCCC's Paris Agreement target to hold the global temperature increase well below 2°C." Reaching the climate target without nuclear energy would be "very difficult."[12]

Another good thing happened in 2015: the United Nations General Assembly agreed to 17 Sustainable Development Goals (SDGs). These goals provide a range of objectives for stimulating action over 15 years,

---

8. Tom Wright is Professor of New Testament and Early Christianity at the University of St. Andrews, Scotland.

9. UNFCCC, *Adoption of the Paris Agreement*, 4, para. 21.

10. Schleussner et al., "Differential Climate Impacts," 327. See also a short presentation from Lindo, "Differences in Climate Impacts," which uses the work of Schleussner et al. as well as Drijfhout et al., "Catalogue of Abrupt Shifts."

11. Poinssot et al., "Assessment of the Environmental Footprint," 200.

12. *WNN*, "Climate Target 'Very Difficult' Without Nuclear."

focusing upon humanity, our planet, and particularly on its economic, social, and environmental dimensions.[13] Peaceful nuclear technologies can play a role here, states the International Atomic Energy Agency (IAEA), since it includes nuclear power under Goal 7, Affordable and Clean Energy: ensure access to affordable, reliable, sustainable, and modern energy for all. The IAEA supports the efficient, safe, and secure use of nuclear energy for current and new plants.[14]

Further, under Goal 13, Climate Action: take urgent action to combat climate change and its impacts: "The IAEA works to increase global awareness of the role of nuclear power in relation to climate change, in particular to try to ensure that the role that nuclear power can and does play in assisting countries to reduce their greenhouse gas emissions is properly recognized."[15] Nuclear energy has its real benefits. Moreover, the IAEA highlights the important contribution of nuclear technologies in achieving other goals, e.g., in Goal 3, Good Health and Well-Being: ensure healthy lives and promote well-being for all at all ages. Here they explain the contribution of research reactors ("that produce life-saving radioisotopes"), radiation oncology, and nuclear medicine.[16] The SDGs are consistent with Christian mission, showing love to God and our neighbors.[17] But what do we need besides technologies? We need sufficient numbers of skilled people.

Skills Shortages

Encouraging people into science is a challenge. Emeritus Professor of Biochemistry Andrew Halestrap (University of Bristol) is a scientist, Christian, and National Chair of Christians in Science (CiS). His CiS leaflet encourages Christians to consider a career in science as a Christian vocation: "it is a fulfilment of the Biblical command in Genesis 1:28 to be good stewards of God's world."[18]

13. UN, "Sustainable Development Goals."
14. IAEA, *Atoms for Peace and Development*, 6.
15. Ibid., 8.
16. Ibid., 4. The World Nuclear Association also recognized the important contribution nuclear energy can make to the SDGs, as reported in *WNN*, "Nuclear 'Vital,'" September 2016.
17. In March 2016, the evangelical John Ray Initiative ran a conference on the SDGs: A Sustainable Future? For the conference resources see http://www.jri.org.uk/events/a-sustainable-future-jri-conference-5-march-2016/.
18. Halestrap, *Thinking About . . . Science as a Christian Vocation*.

Without the right people, we cannot safely and securely operate our energy infrastructure, and there is concern over a nuclear skills shortage. For example, the UK's nuclear industry requires trained staff to operate, maintain, and decommission nuclear power plants. It requires staff for research and development.

New nuclear plants require qualified and experienced staff, but worldwide completion is "fierce."[19] For example, in February 2016 the American Health Physics Society's *Health Physics News* reported on the severe skills shortages identified in the National Council on Radiation Protection and Measurements (NCRP) statement *Where Are the Radiation Professionals (WARP)?*[20] The NCRP commented that the US "is on the verge of a severe shortfall of radiation professionals such that urgent national needs will not be met."[21] Its conclusion was clear: "The looming shortage of radiation professionals represents a serious threat to the United States: scientific leadership is being lost, competition in world markets is affected, and protection of our citizens and country diminished."[22]

In the UK, concern over skills shortages has focused the nuclear industry and government. The company Cogent analyzed the current and future skills needs of the UK's 44,000 civil nuclear industry workforce, concluding that there is "an ageing workforce driving replacement demand; a shift in skills to decommissioning; and, new demand for skills to operate a new fleet of nuclear power stations."[23]

Interestingly, a Nuclear Industry Association (NIA) 2014 survey reported that 35 percent of students studying STEM subjects (science, technology, engineering, and mathematics) did not believe working in nuclear was an option for them.[24] The NIA is tackling this with a campaign to highlight the number of quality careers available, particularly as up to 140,000 workers could be needed before 2030 for the new nuclear plants.[25] Informing students of these opportunities produced positive responses.

19. HLSCST, *Nuclear Research and Development*, 49.
20. Kirner, "WARP Report," 4.
21. NCRP, *Where Are the Radiation Professionals?*, 1.
22. Ibid., 4.
23. Murphy et al., *Power People*, 49. It looked at the workforce requirements from 2009 to 2025.
24. NI, "Nuclear Organisations Join Forces."
25. Ibid.

A 2015 article in *Nuclear Future* asked how educators inspired children to study the STEM subjects and choose careers in "nuclear science, technology and engineering."[26] It draws upon Professor Louise Archer and her colleagues' research which shows children aged 10 or 11 can enjoy science but then do not necessarily want to choose a science career. Crucially, children need to be introduced to STEM subjects before secondary school.[27] The report concluded:

> It is commonly agreed that more needs to be done to improve (widen and increase) STEM participation. The science workforce remains insufficiently diverse—and particular fields are predicting, or currently experiencing, STEM skills gaps that will impact negatively on the economy. There is also a pressing need to improve the spread of scientific literacy across all social groups in society.[28]

Citizens should be informed and involved—ignorance and inaction are not wise choices. On diversity, the Nuclear Institute supports the education of children, a Young Generation Network (YGN), and Women in Nuclear UK (WiN UK).[29]

There are many opportunities in the industry for apprenticeships, degrees, and degree apprenticeships. Nevertheless, as Pollard and colleagues observed, the number of students awarded UK first and higher degrees in STEM subjects rose in the years 2005–2014, but as a proportion of all qualified students in all subjects this represents a fall over the same period.[30]

An article in the UK's *Journal of Radiological Protection* noted: "By 2025, 70% of the UK's existing nuclear workforce will have retired. At the same time, the strong demand for persons with radiation protection knowledge and skills in the non-nuclear and health care sectors is unlikely to diminish."[31] The Society for Radiological Protection's strategy is to edu-

---

26. Olajide et al., "Challenging Perceptions of STEM," 46. Olajide has an engineering degree and works with his coauthors for the Smallpeice Trust (http://www.smallpeice-trust.org.uk) in collaboration with the National Nuclear Laboratory.

27. Archer et al., *Ten Science Facts & Fictions*.

28. Archer et al., *ASPIRES* (2013), 27. Professor Louise Archer is director of the ASPIRES research team.

29. See http://www.nuclearinst.com/YGN and http://www.nuclearinst.com/women-in-nuclear. WiN UK has a charter (http://www.nuclearinst.com/WiN-Industry-charter).

30. Pollard et al., "Taking Nuclear Skills Development." At the time of publishing this article (early 2016) Clare Pollard and coauthors worked for the nuclear group AREVA RMC.

31. Cole et al., "Strategies for Engaging," N26. This 2015 article was coauthored by

cate secondary school pupils in the science of radiation protection and encourage them to consider a career in the profession.[32]

Opportunities also exist for those without STEM subject degrees. For instance, Anna Duckworth graduated from Liverpool Hope University with a degree in Education Studies and English Language. After a while she joined National Nuclear Laboratory (NNL) as a Human Resources Business Partner—read her interview.[33] EDF Energy employs people with many different skills, e.g., at the Hinkley Point C nuclear plant currently under construction.[34]

On another positive note, *World Nuclear News* (September 2016) reported a rise in UK civil nuclear jobs by 2,000 over the previous year.[35] And in December 2016 the UK's Nuclear Skills Strategy Group (NSSG) produced a substantial report, *Nuclear Skills Strategic Plan*, on building excellence in nuclear skills. Then the UK government's green paper (January 2017) on industrial strategy post-Brexit identified many key issues, including skill shortages in (1) basic skills; (2) high-skilled technicians below graduate level; (3) sectors that depend on science, technology, engineering and math (STEM); and (4) sector specific shortages, e.g., the nuclear industry. Finally, the green paper identifies the issue of lifelong learning with the need to up-skill and re-skill people.[36] The government describes its approach to meet the challenges. So clearly, skill shortages are not unique to nuclear.[37] Let us look at other future issues.

---

15 professionals. P. Cole is the Radiation Protection Officer at Liverpool University, and coauthor B.T. Gornall is the Media Officer for the Society for Radiological Protection (SRP). SRP, founded in 1963 and granted Royal Charter in 2007, includes about 38 percent of its membership from the nuclear sector.

32. Also, EDF Energy is working with school children and teachers. See EDF Energy, "Inspiring Young Minds."

33. Duckworth, "Q&A with Anna Duckworth," 3–4.

34. For example, see the video EDF Energy, "Meet the People."

35. *WNN*, "UK Civil Nuclear Job Count Rises."

36. BEIS, *Building our Industrial Strategy*. See the section "Developing skills," 37–49. The green paper is a consultative document for establishing a long-term strategy. People were invited to respond and shape the process.

37. NSSG, *Nuclear Skills Strategic Plan*. Note that "The Nuclear Skills Strategy Group is an industry-led strategic group working with employers, government and trade unions" (3).

Radioactive Waste and Research into the Fuel Cycle

Chapter 6 considered radioactive waste. We shall say a little more here. The academic Behnam Taebi has addressed moral, ethical, and justice issues related to nuclear energy and the potential for harm its waste can cause to future generations.[38] Taebi rightly links the issue of using uranium (a finite resource) with how used uranium (spent fuel) is processed in its fuel cycle and the resulting radioactive waste. He sees basically two fuel cycles operating globally for disposal of the radioactive spent fuel: the open cycle and the closed cycle.

In the open cycle the spent fuel is treated as radioactive waste and disposed of (to a geological disposal facility, i.e., underground). The closed cycle reprocesses the spent fuel to remove useful plutonium and uranium before it disposes of the remainder as radioactive waste. Taebi argues for a new fuel cycle called "partitioning and transmutation" to reduce the lifetime of the radioactive waste and limit the burden on future generations. His position has considerable merit, linking issues of ethics, energy, and the environment.

Indeed, the UK government's *Nuclear Energy Research and Development Roadmap* (2013) evaluated three future options in nuclear waste: (1) Baseline Pathway, (2) Open Fuel Cycle Pathway, and (3) (transition to) Closed Fuel Cycle Pathway.[39] The Baseline Pathway excludes more nuclear plants being built by 2025, but assumes the UK retains its current nuclear fleet. In the Open Fuel Cycle Pathway, new nuclear plants are built but the spent fuel is disposed of without treatment. This pathway is dependent upon higher supplies of fresh nuclear fuel. But a Closed Fuel Cycle Pathway means that unused uranium-235 and plutonium are recovered and recycled in reactors to generate more electricity. By using the recycling program less fresh uranium fuel is needed. Whichever route is followed, all three pathways require research and development, providing opportunities for employees.

What then is the new fuel cycle, "partitioning and transmutation," mentioned earlier? This is part of advanced fuel processing where "partitioning" refers to separation of particular long-lived radioisotopes from the

---

38. Taebi, "Morally Desirable Option," 169. Behnam Taebi has a PhD in the philosophy of technology and is Assistant Professor of Philosophy at Delft University of Technology in the Netherlands. His expertise includes energy and nuclear ethics. See http://ethicsandtechnology.eu/member/behnam_taebi/.

39. BIS and DECC, *Nuclear Energy Research*, 3–4. There is a diagram of the options on page 4.

spent fuel. These are then transmuted into shorter-lived radioisotopes, e.g., in a fast reactor (see below).[40]

An informative paper by Richard Stainsby noted that there is a compelling case for reuse.[41] He showed how much potential fuel is wasted if spent fuel is not recycled.[42] Removing plutonium from spent fuel and other elements (called minor actinides) reduces the radiotoxicity of the waste much more quickly.[43] Spent fuel with direct disposal will take around 300,000 years to decay to the same radiotoxicity level of mined uranium ore (i.e., to the same hazard). But with removal of plutonium and some other products, this timescale is considerably reduced to around 300 years.[44] In December 2015 Russian institutes announced the successful separation of the minor actinides americium and curium from each other, and from the fission products in spent nuclear fuel.[45] So this is good progress.

Extracted plutonium can be mixed with uranium to produce a mixed oxide (MOX) fuel which can be used in reactors. This extends the availability of nuclear fission reactors by not depleting our natural uranium resources so quickly. However, having fast reactors which fission the largely unused uranium-238 in our current reactors means uranium fuel is potentially available for millennia.[46]

Further, a 2014 article by Poinssot and colleagues examined the environmental impacts of closed and open fuel cycles for reactors. The authors conducted a life cycle analysis of the current French nuclear reactors

---

40. Ibid., 82, Appendix D, "Advanced Fuel Cycles."

41. Stainsby, "Advanced Fuel Cycle," 32. Dr. Stainsby is a top scientist at the UK's National Nuclear Laboratory.

42. For a pressurized water reactor with a typical spent fuel assembly, he shows that only 15 kg of the original 20 kg of U-235 is consumed, leaving 5 kg U-235 still available. Moreover, 5 kg of plutonium-239 has been produced from the capture of neutrons in U-238. See ibid., 32–33.

43. Americium, neptunium, and curium are called minor actinides (an actinide has an atomic number in the range 89–103, with actinium at 89). They are formed by neutron capture in uranium-238 followed by radioactive decay, etc. Americum-241 is used as a radioactive source in smoke detectors in homes, thus saving many lives.

44. Stainsby, "Advanced Fuel Cycle," 33. Fig. 2 is helpful.

45. *WNN*, "Russia Hails Progress with Americium."

46. Ibid. Stainsby sees the use of MOX fuel in thermal fission reactors producing a lifetime extension of about a century, while using fast reactors gives a much longer extension. Clearly, this depends on assumptions, e.g., the number of reactors worldwide. For fast reactors Stainsby noted: "but estimates of around 3000 years are not unreasonable." Stainsby, "Advanced Fuel Cycle," 33. Does this count as sustainable?

(which have a closed fuel cycle) and compared this with an equivalent open fuel cycle or once-through fuel cycle (OTC).[47] The OTC considers all spent fuel as waste while the French twice-through fuel cycle (TTC) recovers and reuses plutonium and uranium.

On proliferation, the authors commented: "The French policy aims to strictly balance the plutonium inventory which is recovered by the reprocessing with the one used to produce MOX fuel, so that no stockpile of pure plutonium is accumulated."[48] France's fuel enrichment plant uses nuclear electricity powered by its onsite pressurized water reactors (PWRs) and so its greenhouse gas emissions are lower than other enrichment plants.[49]

The article concluded that, for the same energy produced, an open fuel cycle compared unfavorably with the French twice-through cycle:

- has a larger environmental footprint for "non-radioactive indicators" (e.g., GHGs);
- would give much more high-level waste for the repository (a scarce resource)
- uses 17 percent more natural uranium fuel.[50]

One small benefit of the open cycle was producing lower radioactive discharges. But these discharges were small. Therefore, the French twice-through fuel cycle is superior.

Fast Reactors

The conventional reactors we use, for instance, pressurized water reactors (PWRs), use thermal (or slow) neutrons to achieve fission in the fissile uranium-235 in the nuclear fuel. The neutrons that are given off in fission arise as fast neutrons. They are slowed down by a moderator (e.g., light water) to give slow neutrons, which cause fission in the uranium-235. But over

---

47. Poinssot et al., "Assessment of the Environmental Footprint," 199. The article clearly explains the fuel cycle.

48. Ibid., 201.

49. Ibid., 201–2. The enrichment plant increases the amount of uranium-235 in the fuel from the natural level of 0.7 percent when mined to the few percent needed in reactor fuel. France strategically chose to expand its nuclear capacity after the 1973 oil crisis and this capacity allows it to use nuclear generated electricity for its enrichment process. For the French nuclear fuel cycle see ibid., fig. 1.

50. Ibid., 210. The repository is an underground waste disposal/storage facility.

90 percent of the fuel is uranium-238 and this does not normally undergo fission. However, fast reactors produce fission in the uranium-238, thus recovering much more energy from the fuel.

Demonstration fast reactors operated for decades in some countries before being closed down (in the UK the government withdrew its funding for fast reactors).[51] But now there is a worldwide interest in developing them for commercial operation. For instance, in August 2016 the Russian Beloyarsk-4 BN-800 fast reactor achieved 100 percent power (it had previously reached lower power levels and generated electrical power). It entered commercial electricity generation in November 2016.[52]

"Too Cheap to Meter"

"Too cheap to meter" is frequently quoted to illustrate nuclear energy's unfulfilled promise and so question its future role, as Brown and colleagues do.[53] But is it accurate? If a nuclear plant is constructed the initial capital costs are high (as people recognize), fuel has to be purchased, the plant must be maintained, staff have to be paid, etc. All this clearly costs, so how could anyone realistically claim nuclear energy would be "too cheap to meter", unless, by that you mean that consumers pay a fixed fee irrespective of the amount used?[54]

A 2009 article by M. J. Brown asked: "Too Cheap to Meter?" It noted that Lewis L. Strauss (Chairman of the US Atomic Energy Commission) did mention in speeches (1954 and 1955) his "futuristic vision" that children will have homes with *"electrical energy too cheap to meter."* But his precise meaning is unclear and disputed. Brown remarked:

---

51. Hore-Lacy, *Nuclear Energy*, 49. This included over 30 years' operation in the UK. Fast reactors still operate, for example, in India, Japan, and Russia. Also see WNA, "Generation IV Nuclear Reactors," and "Fast Neutron Reactors."

52. WNN, "Russian Fast Reactor." The BN-800 had previously been connected to the grid in December 2015. It is a 789 MWe reactor using fuel with uranium and plutonium oxide. Also, see WNN, "Russia's BN-800 Unit Enters."

53. Brown et al., *Great Transition*, 53, stated: "Nuclear energy, once lauded as an energy source that would be 'too cheap to meter,' is becoming too costly to use." When the 2015 book was published the authors were with the Earth Policy Institute. However, Lester Brown (president and founder), aged 81, stepped down and the institute closed in June 2015 (http://www.earth-policy.org).

54. Many UK homes do not have water meters, but that does not mean it is free! They are charged a fixed rate irrespective of the volume of water used.

Amazingly, over fifty years later one still hears the phrase *"too cheap to meter"* repeated by the media and the antinuclear movement, construed as a proof that nuclear science and technology promised some Utopian future but never delivered. But did the nuclear industry really expect to generate electricity *"too cheap to meter"*? Or is it just an easy, catchy and tired phrase taken out of context?[55]

Important voices around that period demonstrated that this was not the widespread view on costs. Indeed, for the UK, Brown quotes Sir John Cockcroft on the UK's reactor development: *"we do not expect to produce a cheaper source of power than that derived from coal—it is likely, in fact, to be somewhat more expensive. What we are aiming at is to increase the total power available."*[56] Therefore, Brown concluded that the nuclear industry saw a future for nuclear, but realized the practical costs didn't propagate a vision of "too cheap to meter." Strauss's vision was futuristic, but his phrase was disputed: it did not represent the industry.

Moreover, K. S. Parthasarathy's article "The Fallacy of Nuclear Power Being Too Cheap to Meter" supports the view of M. J. Brown (in his Canadian Nuclear Society article) that the phrase is usually taken out of context.[57] Finally, Robert Deis concludes that wherever we stand in the nuclear debate, through whatever energy source we generate electricity, the popular saying is true: "There's no such thing as a free lunch."[58]

Nuclear Energy Post-Fukushima

We have already discussed accidents in chapter 6, including the one in 2011 at Fukushima, but here I wish to ask how this has impacted on nuclear energy. First, a 2015 Pew Research Center report found that 65 percent of scientists in the American Association for the Advancement of Science who were surveyed supported nuclear new builds, higher than the 45 percent of citizens.[59]

55. M. J. Brown, "Too Cheap to Meter?" Italics original.
56. Ibid. Stated in the 1951 Joule Memorial Lecture. Italics original.
57. Dr. K. S. Parthasarathy is a former secretary in the Atomic Energy Regulatory Board, India.
58. Deis, "'Too Cheap to Meter," September 17, 2015. Also, Deis calls it "the infamous nuclear power misquote."
59. Funk and Raine, *Public and Scientists' Views*. Also, 87 percent of the scientists accepted that climate change was mainly human induced, but only 50 percent of citizens accepted this.

Further, an important article in the journal *Renewable and Sustainable Energy Reviews* evaluated the impact of the nuclear accident in Fukushima on government nuclear policies and development worldwide, including those in China.[60] It examined the impact immediately after the accident and then more recently. For example, Germany decided to permanently shut down some reactors and phase out the remainder, but most countries chose to continue with the nuclear option.[61] China, after an initial halt in development, has accelerated its development of nuclear power.[62]

In 2016 New York approved a Clean Energy Standard that recognized the valuable contribution that its local nuclear power plants made to carbon-free electricity generation and climate change targets.[63] Also in 2016, Illinois passed its Future Energy Jobs bill to secure ongoing operation of its at-risk nuclear plants (Clinton and Quad Cities) by recognizing their contribution to zero-carbon emissions.[64] Again in 2016, Wisconsin's governor signed a bill that ended the state's moratorium on the construction of new nuclear plants.[65] However, much more remains to be achieved in the US to preserve its nuclear plants. To this end, Daniel Shea and Kristy Hartman authored for the bipartisan organization the National Conference of State Legislatures (NCSL) the valuable report, *State Options to Keep Nuclear in the Energy Mix* (January 2017). This provides case studies on the bills, policy options, and possible strategies that inform state legislatures on keeping nuclear plants from premature closure.[66] For a short report on the worrying premature shutdown of US reactors and why people should know this, and also care, see Adams' article "What's Really Killing America's Nuclear Plants." He regrets that the problem rests with the nuclear energy industry,

---

60. Ming et al., "Nuclear Energy in the Post-Fukushima Era." Zen Ming and most coauthors are from the School of Economics and Management, North China Electric Power University, Beijing, and one author is from Shenzhen Power Supply Co., Shenzhen.

61. Ibid., 148–49.

62. Ibid., 153. For a further discussion see: OECD and NEA, *Impacts of the Fukushima Daiichi Accident* (March 2017). The report addresses other factors that impacted on energy policies besides the Fukushima accident, and these factors may have had more important influences.

63. *WNN*, "New York Approves." Also, see NEI, "New York's Support for Nuclear."

64. *WNN*, "Illinois Energy Bill." See NEI, "Map of US Nuclear Plants," Illinois.

65. *WNN*, "Wisconsin Lifts."

66. Shea and Hartman, *State Options*.

which "has not invested enough in telling people why they should value this important technology."[67]

A Future with Renewables and Reactors

My argument, shared by many people, is that we need both renewables and nuclear energy. A paper by Ted Trainer which considered energy needs in 2050, and sought to keep within safe levels of greenhouse emissions, asked: "Can renewables etc. solve the greenhouse problem?" It answered No. The author's analysis assumed nuclear was present at the current level of generation, but he argued that even with coal and hydroelectricity it cannot meet supply gaps when there is little or no output from solar and wind (e.g., in winter). Considerable changes to lower economic production and consumption are required, together with changes in "social structures and systems."[68]

Nevertheless, Australian scientist Barry Brook reckoned the analysis neglected a major increase in nuclear electricity generation because of assumptions on (1) currently assessable uranium reserves and (2) life-cycle emissions. Rather, Brook observed: "on technical and economic grounds, nuclear fission could play a major role (in combination with likely significant expansion in renewables) in future stationary and transportation energy supply, thereby solving the greenhouse gas mitigation problem."[69] Nevertheless, Brook rightly saw the challenge of society's acceptance, and public education, to accept nuclear's expanding role as a principal limitation.[70]

The International Energy Agency's report *World Energy Investment 2016* states that various issues hinder investment in nuclear energy both in Europe and the US. In the European Union (EU) decommissioning of nuclear plants and fossil fuel plants has increased concerns on energy security, so the IEA report commented: "In several countries, nuclear capacity is

---

67. J. Adams, "What's Really Killing." Adams is founder and CEO of Full On Communications.

68. Trainer, "Can Renewables etc. Solve," 4107. Ted Trainer's affiliation when his 2010 article was published was in Social Work, University of NSW, Australia.

69. Brook, "Could Nuclear Fission Energy," 4. When he wrote this 2011 article Brook was at the Centre for Energy Technology and School of Earth and Environmental Sciences, University of Adelaide, Australia.

70. Ibid., 4, 7. He includes fossil fuels with carbon capture and storage in his energy mix, considerable electrification, and an expansion of nuclear energy. Other principal limitations are "fiscal and political inertia" and "inadequate critical evaluation of the alternatives."

ageing with little investment going to replacement capacity, and renewables are struggling to compensate for reduced nuclear output."[71]

In June 2016 at the second G20 energy ministers' meeting (in China) the energy ministers reaffirmed their commitment to energy that is affordable, reliable, and sustainable by issuing a communiqué.[72] Did this only address renewables? Well, no. For countries using nuclear energy it rightly encouraged attention to the highest standards of: safety, security, nonproliferation, regulation, exchange of expertise/experience, and public dialogue which is science based.[73] The EU too promotes "the highest standards of nuclear safety across Europe and beyond," while recognizing that 30 percent of the EU's electricity is nuclear generated.[74] Moreover, the European Commission's short report (*Nuclear Illustrative Programme*) on investments in nuclear safety focused on post-Fukushima upgrades to plant and safe operation.[75] It is a mine of information for stakeholders and the public. For example, is nuclear declining in the EU? Well, between now and 2025 the EC predicts a decline in total EU nuclear capacity. But this trend will be reversed by 2030 with new reactors coming online and reactor lifetime extensions. From 2030 to 2050 the nuclear capacity remains fairly stable.[76]

Nevertheless, the EC's *Nuclear Illustrative Programme* was evaluated by the European Economic and Social Committee (EESC) and it found a number of shortcomings in its strategy that required "substantial revisions." *World Nuclear News* reported the EESC's view that "the commission should highlight nuclear energy's positive attributes."[77] In 2017, writing in the *Bul-*

---

71. NucNet, "'Major Shift' Has Begun." NucNet is an independent global nuclear news agency. Also, for the executive summary, with the above citation, see Starling, "IEA: World Energy Investment 2016."

72. EC, "G20 Energy Ministers Commit."

73. G20, *Energy Ministerial Meeting*, "Nuclear Power."

74. EC, "Nuclear Energy."

75. EC, "Commission Presents Report."

76. EC, *Nuclear Illustrative Programme*, 4. However, as electricity demand increases over the period, the percentage share of nuclear falls from the current level of 27 percent to about 20 percent.

77. WNN, "Europe Needs to Revise." Further, in November 2016 the European Atomic Forum (FORATOM) stated that the *Nuclear Illustrative Programme* offered a "snap shot" with many positive aspects (e.g., in Europe 27 percent of electricity is from nuclear generation and 27 percent from renewables), but "what is now needed is a vision and strong leadership in order to promote nuclear as part of the solution to climate change." See FORATOM, *FORATOM's Further Commentary*, 1. FORATOM is Europe's nuclear industry trade association.

letin of the Atomic Scientists, the academic Michael M. May sensibly stated: "Nuclear energy, hydroelectric power, and renewables are all needed to meet the goals of the 2015 Paris Agreement on climate change, and the 10 largest emitters of greenhouse gases all plan to use nuclear power in some way to deal with the climate crisis."[78] This agrees with my view that a more positive nuclear narrative is needed (a new nuclear narrative).[79]

As an illustration of renewables and reactors supplying low-carbon energy, let's look at the UK. In February 2017 the UK's department BEIS published preliminary data that showed the share of electricity generation from major power producers in 2016. Low-carbon generation amounted to 42.7 percent (42.9 percent in 2015) of supply. Nuclear energy accounted for 23.8 percent (23.0 percent in 2015). Wind contributed less generation than in 2015 (as average wind speed was 11 percent lower), and hydro also contributed less in 2016 following a reduction in average rainfall of 19 percent. Outages (i.e., shutdowns) at Drax Power Station meant less bioenergy contribution to supply.[80]

In April 2017, the European Commission published its package *Clean Energy for All Europeans*. However, FORATOM, the Brussels-based European nuclear trade association, has reacted to it by welcoming some aspects but criticizing others. FORATOM's position paper "considers that the package, including as it does EU governance of energy and climate policies, could ensure a coherent and optimal approach towards meeting the EU 2030 energy and climate objectives, if it takes into account the views of the nuclear industry."[81] FORATOM sees nuclear energy as essential to achieving "the EU's energy and climate goals" and offers its proposals.

Also in April 2017, *World Nuclear News* reported that the French Academy of Sciences had highlighted contradictions in France's policy bill Energy Transition for Green Growth (approved July 2016). The bill's objectives are: reductions in greenhouse gas emissions, reduction in overall energy consumption, increased renewable energy but reduced nuclear

---

78. M. M. May, "Safety First," 38.

79. Steve Kidd has argued for the nuclear industry to engage in a "paradigm shift" that moves people away from "nuclear fear" or the "fear paradigm," rather than just offering people more facts and figures. See Kidd, "Nuclear Establishment." He is an independent nuclear consultant and economist working for East Cliff Consulting.

80. BEIS, *UK Energy Statistics—2016 Provisional Data*, 3–4. The more detailed analysis was published as BEIS, *Energy Trends March 2017*. It has a short article called "Special Feature – Nuclear Capacity in the UK," 92–94.

81. FORATOM, "Reaction," 1.

energy's share in electricity generation (from around 75 percent currently to 50 percent by 2025). Nevertheless, France's Academy of Sciences saw the debate on energy transition as incomplete. It considered that energy policy programs needed to be realistic, evaluating "physical, technological and economic constraints." Instead, however, "In the current debate, our fellow citizens might be led to believe it would be possible to develop renewable energy as a way to decarbonise the system while removing both fossil fuels and nuclear." It explains some shortcomings of renewables (e.g., their intermittent availability) The Academy saw a contradiction in trying "to reduce emissions whilst reducing the share of nuclear power."[82]

Thorium Reactors

In some countries thorium-fueled reactors are being developed, since thorium is a more abundant element in the Earth than uranium, and has other advantages.[83] For example, China and India both have substantial thorium supplies. Many reactors are suitable for fueling with thorium and a number of thorium-fueled demonstration reactors have run over the years, e.g., in the US.[84]

The UK's House of Commons Energy and Climate Change Committee (HCECCC) evaluated the thorium option, with its pros and cons, but the Committee saw advantages in a thorium fuel cycle. Nevertheless, they thought the option would not be viable unless uranium fuel costs changed considerably.[85] Still, it recommended continued research and development, with the government examining potential long-term benefits and potential barriers. So, the thorium fuel option remains open.

---

82. *WNN*, "Academy Highlights Contradiction."

83. Hore-Lacy, *Nuclear Energy*, 35, noted thorium is around three times more abundant than uranium.

84. WNA, "Thorium." Also, in Norway's Halden research reactor, thorium is being tested as an additive to uranium fuel (called Th-Add) for use in existing reactors. Further, fuel rods of thorium mixed with plutonium (called Th-MOX) are being irradiated in a trial. See in ibid. the subsection "Thorium as a Nuclear Fuel."

85. HCECCC, *Small Nuclear Power*, 9. Written evidence was submitted to the committee by the All Party Parliamentary Group on Thorium Energy.

Small Modular Reactors

Small modular reactors (SMRs) are potential future reactors. They are small, scalable low-carbon reactors with many advantages over the much larger reactors that are currently built to generate electricity. Moreover, interest in evaluating, developing, licensing, and operating SMRs is not confined to the UK, but there is international interest.[86] SMRs could be used to generate electricity, power seawater desalination plants, provide district heating, etc.

But what are they? They are small reactors usually under 300 MWe, so much smaller than those proposed for the UK's new nuclear build (e.g., Hinkley Point C is 1,600 MWe per reactor).[87] Over the years, reactors in the UK were built larger (the early Magnox reactors were smaller than the later ones) as this gave economies of scale. However, larger reactors have disadvantages such as high initial construction costs. Industry and government have expressed an interest in investigating the viability of SMRs in the UK and the possible overseas markets for the UK to enter.

The UK's National Nuclear Laboratory (NNL) published in 2014 a *Small Modular Reactors (SMR) Feasibility Study* outlining their advantages.[88] Nonetheless, NNL recognized that no SMRs were operating and challenges—financial and regulatory—beyond technical aspects were still to be met.

Also in 2014, a House of Commons Energy and Climate Change Committee (HCECCC) report considered SMRs in more detail and received evidence from individuals and organizations.[89] Its conclusions and recommendations addressed: the potential role of small nuclear power, technologies and fuels, cost and investment risk, regulatory assessment, siting considerations, safety and security, and public engagement. The conclusions stated for the UK were:

---

86. Clegg and El-Shanawany, "Small Modular Reactors." When they wrote their article, Professor Richard Clegg was the Global Nuclear Director for Lloyd's Register and Professor Mamdouh El-Shanawany was the Nuclear Technical Director of Energy at Lloyd's Register. They noted that SMRs are not new since naval propulsion used them, with around 700 nuclear reactors in submarines, aircraft carriers, plus icebreakers. They estimated 200 currently in use (2012).

87. IAEA, *Status of Small Reactor Designs*, 1, 9. IAEA stated: "small reactors are the reactors with an equivalent electric power less than 300 MW, medium sized reactors are the reactors with an equivalent electric power between 300 and 700 MW." But sometimes reactors up to 500 MWe are considered as small.

88. NNL, *Small Modular Reactors*, 60.

89. HCECCC, *Small Nuclear Power*. Written by a cross-party committee of Conservatives, Labour, and one Liberal Democrat.

- SMRs are viable with a key role for low-carbon energy in the next decade;
- government should be proactive in driving forward development and deployment;
- scope exists for British industry to be involved.[90]

SMRs must undergo the normal regulatory process. However, the committee was surprised that to obtain a nuclear site licence can take up to six years. So, three issues were identified: (1) for the government to try to improve the Generic Design Assessment, (2) to ensure that the Office for Nuclear Regulation (ONR) is suitably resourced, and (3) for the ONR to consider ways of streamlining its regulatory process. The importance of public engagement was rightly addressed.[91]

Moving forward, in 2016 the UK government announced a competition for SMRs.[92] United States company NuScale Power confirmed it would enter the competition. Moreover, in December 2016 the company submitted its application to the US Nuclear Regulatory Commission to review and approve its SMR design for construction and operation.[93] An advantage of NuScale's SMR is its ability to use conventional low-enrichment uranium light water reactor fuel, but it can also operate with mixed uranium-plutonium oxide (MOX) fuel. This would be a valuable way for the UK to use up its long-standing civil plutonium stocks (of over 100 tons).[94] It is a win-win situation with the plutonium removal (thus eliminating a proliferation risk) and carbon-free electricity generated. Of course, NuScale is not the only company interested here; Tim Fox of the Institution of Mechanical Engineers (IMechE) recognized the potential of also achieving this by utilizing the Canadian company Candu Energy and constructing their proven-technology CANDU 6 (EC6) reactors.[95]

---

90. Ibid., 26.

91. Ibid., 25–26. For a further overview, see WNA, "Small Nuclear Power Reactors."

92. DECC, *Small Modular Reactors*. As a reminder, the UK's Department of Energy and Climate Change (DECC) merged into the new Department for Business, Energy and Industrial Strategy (BEIS) in July 2016.

93 *WNN*, "NuScale Makes History." NuScale's SMR application is the first one submitted to the NRC. Its application included almost 12,000 technical pages and it will submit further information if requested by the NRC.

94. *WNN*, "Study Confirms NuScale." The independent study was made by the UK's NNL. See also NuScale Power, "NuScale Announces MOX Capability."

95. Fox, "UK Plutonium." Dr. Tim Fox was then Head of Energy and Environment at the IMechE. Westinghouse Electric Company was also interested in building UK SMRs.

In October 2016 the UK's Rolls-Royce indicated its interest in entering the SMR market and experts have called on the UK government to clarify its requirements.[96] Subsequently, Rolls-Royce named some UK partners and submitted a paper to BEIS outlining plans for building SMRs in the UK.[97] In December 2016 the government responded to the House of Commons Energy and Climate Change Committee (HCECCC) October 2016 report on *The Energy Revolution and Future Challenges* and indicated its ongoing support for nuclear power and its recognition of the potential benefits of SMRs. Sensibly, it required that a "robust evidence base" was needed to evaluate SMRs and their deployment.[98] And in March 2017 *World Nuclear News* reported "UK Businesses Plan for Global SMR Market." However, the House of Lords report, mentioned previously, was published in May 2017 which provided support, recommendations, and concerns on SMRs. For example, the House of Lords Science and Technology Committee stated "The Government's failure to make a decision on its strategy for SMRs is a prime example of its inaction in the civil nuclear arena. Not keeping to the stated timetable for the competition has had a negative effect on the nuclear sector in the UK . . . "[99] Whatever the outcome from the competition, any construction will provide employment and economic benefits, but we'll need to see how all this develops.

However, in March 2017 Westinghouse filed for Chapter 11 bankruptcy in the US to protect its interests from creditors while it completed "strategic restructuring" related to its US AP1000 power plants. See *WNN*, "Westinghouse Files." We'll need to see how this develops.

96. *WNN*, "UK Considers."

97. *WNN*, "Rolls-Royce Names Partners." These companies regard SMRs as complimentary to the program of building large reactors in the UK.

98. HCECCC, *Energy Revolution.* BEIS, *Fourth Special Report—Appendix: Government Response*, Recommendation 8. Committee recommendations are in bold and Government responses in plain text. Also, see the helpful discussion in *WNN*, "Nuclear's Role."

99. HLSTC, *Nuclear Research and Technology*, 4. See ch. 4, "Small Modular Reactors." Also, see NIRAB, *NIRAB Final Report: 2014-16*. Recommendation 10 states "Government should make clear its aims for SMR development in the UK, ensuring that these are used in evaluating the SMR competition," (v).

Fusion Reactors

Commercial operation of fusion reactors may be far off, but there is considerable ongoing research and development.[100] Fusion is the way energy is generated in our sun (see chapter 1). Let me mention a few things.

First, the UK 2013 report *Nuclear Industrial Strategy* stated:

> The UK has a recognised world leading status in fusion research ... CCFE [Culham Centre for Fusion Energy] is the main centre of UK fusion research with the operation of JET (Joint European Torus), the only device in the world capable of fusion, under contract from EURATOM. Success here has positioned the UK to be a major participant in the operation of the International Thermonuclear Experimental Reactor (ITER), under construction in the south of France. The EU roadmap projects fusion to generate electricity in the 2040s.[101]

Second, the UK's Engineering and Physical Sciences Research Council (EPSRC) has independently reviewed fission and fusion research, giving major findings and recommendations. Considerable support is provided for the direction of further research in fission and fusion reactors.[102]

Third, Christians are involved in research on fusion reactors. For example, Ian Hutchinson became a committed Christian while an undergraduate physics student at Cambridge University. Now he is Professor of Nuclear Science and Engineering at Massachusetts Institute of Technology (MIT) in the US, where he researches, with his team, plasma physics and nuclear fusion using an international facility. Their work contributes to the development of a future fusion reactor.[103]

Commercial fusion reactors are some time off in the future, but let us recap the current benefits from nuclear energy.

---

100. Christian author Ian Hore-Lacy outlines nuclear fusion, including its pros and cons, in Hore-Lacy, *Nuclear Energy*, 18. Fusion reactors will give us carbon-free electricity, but the capital costs are expected to be high and some radioactive waste will be generated. Also, see WNA, "Nuclear Fusion Power." There are still scientific and engineering issues to resolve.

101. BIS, *Nuclear Industrial Strategy*, 65.

102. EPSRC, *EPSRC Independent Review*.

103. Hutchinson and Anderson, "Being a Christian and a Scientist."

Let There Be Light!

Nuclear Energy: The Benefits

An International Atomic Energy Agency 2015 report stated:

> Nuclear power can make a significant contribution to reducing greenhouse gas emissions while delivering energy in the increasingly large quantities needed for growing populations and socio-economic development. Nuclear power plants produce virtually no greenhouse gas emissions or air pollutants during their operation and only very low emissions over their entire life cycle. Nuclear power fosters energy supply security and industrial development by providing electricity reliably at stable and foreseeable prices.[104]

As a specific UK example, on February 5, 2016, EDF Energy's Hinkley Point B nuclear plant celebrated 40 years of safe and reliable electricity production since it was first connected to the national grid in 1976 (I was working there then). EDF Energy reported that the two reactors had "avoided the production of around 187 million tonnes of carbon dioxide and greenhouse gases—the equivalent of taking all passenger cars off UK roads for just under two and a half years."[105] This significant achievement should be appreciated. Furthermore, the Nuclear Industry Association (NIA), which represents Britain's civil nuclear industry, noted: "The power generated by existing power stations avoids the emissions of 49 million tonnes of CO_2 a year—the equivalent of taking 78% of Britain's cars off the roads."[106]

EDF Energy states that the Hinkley Point C reactors (3.2 GWe) will power about 6 million homes and avoid annually around 9 million tons of CO_2 emissions.[107] This is a massive ongoing saving over its 60-year lifespan.

Moreover, nuclear energy contributes to clean air and the reduction of millions of annual premature deaths and diseases from breathing polluted air.

Nuclear provides a constant supply of electricity which is weather independent, thus giving a secure baseload supply. A criticism sometimes leveled against nuclear is its inflexibility, as it cannot adjust its output

---

104. IAEA, *Climate Change and Nuclear Power 2015*, Foreword. An example of air pollution from coal plants is discharges of mercury. See Evangelical Environmental Network, "Statement by Evangelical Leaders on Mercury."

105. EDF Energy, *Hinkley Point B*, 2. The report also noted that the two reactors still provided "enough electricity for approximately one and a half million households."

106. NIA, "NIA Welcomes SMR Deployment Report."

107. EDF Energy, *Hinkley Point C: Building* (July 2016), 9. See more important figures on this page, e.g., £100 million will go into the regional economy during peak construction.

according to customer needs on the electricity grid. However, the World Nuclear Association points out that most pressurized water reactors and new reactor designs will be designed to change their electrical output (load following) in operation, thus providing flexibility.[108]

Although nuclear reactors raise concerns over proliferation, the other side of the coin is that nuclear has helped in the nonproliferation of nuclear weapons through the Megatons to Megawatts program (see chapter 3). Russian uranium from nuclear weapons was used as fuel in US reactors to produce electricity—a win-win situation.[109]

Nuclear plants occupy a smaller surface area than the equivalent electrical energy provided by wind and solar generation.[110] In addition, the volumes of waste are tiny for a 1,000 MWe nuclear plant compared with the equivalent coal plant, which releases uncontrolled into the atmosphere 7 million tons of CO_2 annually, plus radioactive fly ash and heavy metals.[111]

Nuclear energy brings economic opportunities into local areas in the UK, as very many people are employed and new people move into the area for high-quality well-paid jobs. Supply chains across the country also provide training and employment opportunities. The UK's Office for National Statistics has identified the vital contribution of nuclear energy to the economy.[112]

As the UK decarbonizes, for example by electrifying its railways, nuclear can help meet this need for more electricity. Of course, reactors have already been used to power submarines and ships. For instance, in 2013 the Olympic torch for the Winter Olympics in Sochi, Russia, was taken to the North Pole in a Russian nuclear-powered icebreaker.[113]

Nuclear power in the UK is regulated by the Office for Nuclear Regulation, which holds the nuclear industry to account on behalf of the public.

108. WNA, "Advanced Nuclear Power Reactors."

109. For more information on the completed program (2013) and the current program with its disputes, see *WNN*, "Russia Suspends Plutonium Agreement" (October 2016).

110. Brand, *Whole Earth Discipline*, 81. He used Gwyneth Cravens's US example that a 1,000 MWe nuclear plant occupies about a third of a square mile, while wind turbines producing the same energy would occupy more than 200 square miles. The Campaign to Protect Rural England is concerned over land—see Palmer, *Warm and Green*, 31, on land-mass usage implications for biomass, nuclear power stations, wind (onshore/offshore), and solar. For a detailed discussion for the US, see Ausubel, "Renewable and Nuclear Heresies." Jesse Ausubel is currently Director of the Program for the Human Environment at The Rockefeller University, New York. See also Ausubel, *Power Density*.

111. Brand, *Whole Earth Discipline*, 81.

112. *WNN*, "Nuclear Plays 'Vital' Role."

113. *WNN*, "Nuclear Ship Takes Olympic Flame." The ship was the *50 Years of Victory*.

This is an important point to reassure the public that the industry is being effectively regulated.

In the US, Berkman and Murphy produced an important 2015 report addressing *The Nuclear Industry's Contribution to the U.S. Economy*.[114] As the title indicates, they examined the substantial contribution of the nuclear industry to the US economy (e.g., costs, employment, and taxes). Then in December 2016 Metin Celebi and colleagues published *Nuclear Retirement Effects on CO_2 Emissions*. They analyzed energy, economics, impacts of early retirements of nuclear plant on CO_2 emissions, and achieving climate goals, e.g., "The potential vulnerability of some nuclear power plants to premature retirement creates a major threat to the attainment of CO_2 reduction goals."[115] Finally, also in December 2016, Berkman and Murphy produced a specific report on *Pennsylvania Nuclear Power Plants' Contribution to the State Economy*.[116] Here they helpfully provide a detailed analysis of the many benefits from Pennsylvania's nuclear energy. Space does not provide an opportunity to discuss this further, but it is well-worth reading to understand the numerous benefits for the economy and environment associated with a particular US state and its nuclear plants.

China has specific needs for nuclear energy besides the need for clean air and emission reductions to reduce smog. For example, it is developing the ACP100 small modular reactor (SMR) technology, which has the potential to be used on a small and medium grid, provide district heating, provide process heating (for industries with high energy requirements), and power desalination of seawater where there is a shortage of fresh water.[117] Another reason for using nuclear energy is to reduce the coal transport

---

114. Berkman and Murphy, *Nuclear Industry's Contribution*. The report was produced by the Brattle Group (http://www.brattle.com/about) for the UK's Nuclear Matters (http://nuclearmatters.co.uk/). The authors also note that nuclear plants avoid air emissions of sulfur dioxide, nitrogen oxides, and particulate matter.

115. Celebi et al., *Nuclear Retirement Effects*, 1. Again, this report was produced by the Brattle Group, with support from the Nuclear Energy Institute.

116 Berkman and Murphy, *Pennsylvania Nuclear Power Plants'*. They prepared the report for the Pennsylvania Building and Construction Trades Council, the Chamber of Business and Industry, the Allegheny Conference on Community Development, and the Greater Philadelphia Chamber of Commerce (see the unnumbered page after the title page).

117. Song et al., "Small Is Beautiful," 45. They expect the first ACP100 construction to begin in December 2017. Its reactor module is underground to protect it from aircraft, terrorism, and external events.

demand. Coal is unevenly distributed across China and it needs to be transported. This has put pressure on its transport system.[118]

We should remember that new reactor developments and designs provide enhanced safety features and security (e.g., against aircraft crashes), including lessons learnt and applied from Fukushima.

Christians, Churches, and Christian Leaders

We need to ask how Christians, churches, and Christian leaders respond to nuclear energy. Specific Christian responses, for instance, include Sir John Houghton's articles and books. His 2009 paper challenged scientists, policy makers, and Christians that as stewards of creation there is an ethical and Christian challenge to global warming.[119] The book *Creation in Crisis*, edited by Robert S. White, has valuable articles on the environment, ethics, science, and the Bible. And Mellen and Hollow's book *No Oil* has two practical chapters.[120]

A 2007–8 Environment Agency (EA) poll with 25 leading environmentalists identified 50 top things to save the planet. Nuclear energy was not specifically mentioned but the second thing was "a leap of faith."[121] The message was that religious leaders must prioritize the planet. On the environment, faith groups needed unity in action rather than continuing silence, while leaders needed to inspire their congregations to care for the planet. Since 2007–8 responses have improved, including the Pope's intervention. Nevertheless, how equipped are Christian leaders to guide their flocks? How knowledgeable are they on the environment, energy options, and particularly on nuclear energy? Thankfully, over the years a number of organizations (including churches and charities), conferences, books, and papers have educated Christians and Christian leaders about these important issues.[122]

---

118. Cheng et al., "HPR1000," 38–39.

119. Houghton, *Global Warming, Climate Change*.

120. Mellen and Hollow, *No Oil*. See their ch. 11, "What Can I Do?," and ch. 12, "What Can My Church Do?"

121. EA, "50 Things," 17.

122. On education and training, see Weaver and Hodson, *Place of Environmental Theology* (related to European Baptists); and Hodson and Rushton, *Faith, Environmental Values* (investigating the training of Anglican ordinands). See also the contributions in White, ed., *Creation in Crisis*.

In recent years, the Anglican Church decided to introduce "Common Awards" at its training colleges for ordinands. Following this church initiative, Durham University became the validating body for their theological qualifications. Bristol Baptist College is also a participant in "Common Awards." My search identified module outlines on the environment. Under individual institutions I selected the regulations for Trinity College with Bristol Baptist College (because of my previous connections with both colleges). An examination of the Durham Awards for the six programs they offer (from Certificate to MA level) showed five modules. However, it is up to individual institutions to choose which modules to offer students from the list; they are not obliged to offer everything.[123] In no programs were these mandatory, since students have the freedom to choose modules from defined lists.[124]

However, Anglican and Baptist ordinands (ministerial candidates) have ministerial pathways that specify mandatory modules, so any required environment modules would be specified there. Thus, it appears that non-ministerial students could obtain a theology award without completing any modules on the environment under these regulations. Moreover, John Bimson has confirmed that Anglican ordinands can complete their theological training without being required to take any modules on the environment.[125] Similarly, Stephen Finamore confirmed the same situation exists for Baptist ministerial students.[126] How much time is devoted to understanding nuclear energy remains unclear without examining the individual ministerial pathways. However, the student handout for the MA module Christian Faith and the Environment shows that nuclear energy is addressed.[127] Moreover, continuing engagement is needed to keep track of current and future policies and developments.

---

123. Durham University, "Individual Module Outlines." The modules are: Christian Faith and the Environment (TMM41820); Environmental Theology (TMM3217); Issues in Science and Religion (TMM2721); Justice, Environment and Mission in Global Context (TMM3441); Science, Ecology and Theology (TMM2661); Sustaining the World: Christian Faith and the Environment (TMM1247); Sustaining the World: Christian Faith and the Environment (TMM2237). There is also a module on Theology and Science (TMM41320).

124. See Durham University, "Programme Documents."

125. John Bimson, Associate Research Fellow Trinity College Bristol, personal communication, December 17, 2016.

126. Stephen Finamore, Principal Bristol Baptist College, personal communication, March 13, 2017.

127. Taught by Dr. John Bimson, formerly full-time staff at Trinity College, Bristol, and a colleague.

Nevertheless, Stephen Finamore makes the important point that Bristol Baptist College has an unaccredited ministerial program that engages with external speakers on the environment, e.g., Friends of the Earth and the Christian organization called A Rocha. Further, environmental issues are addressed in modules not specifically identified as covering the environment. This applies to mission modules and some modules on the Bible. Students cannot cover all issues in their training but they are equipped to approach the world thinking both "theologically and missiologically."[128] These are all valid points.

The worldwide evangelical Lausanne Movement, among many things, has called upon:

A. "Local church leaders to (i) encourage, support and ask questions of church members who are professionally engaged in science, technology, healthcare and public policy, and (ii) to present to theologically thoughtful students the need for Christians to enter these arenas."

B. "Seminaries to engage with these fields in their curricula, so future Church leaders and theological educators develop an informed Christian critique of the new technologies."[129]

There is also a call for theologians and Christians to form partnerships for engaging "with new technologies and scientific advances" so as to speak into public policy with a biblical and relevant voice.[130]

However, how well are Christian leaders able to do this with their knowledge of science? In an article by Rebecca Bouveng and David Wilkinson published in *Science and Christian Belief*, they interviewed senior Christian leaders in the UK on science and faith. Fourteen leaders (Anglican bishops plus leaders in other Christian denominations and organizations) responded to their invitation. They noted: "We ended up with a sample where the majority of the interviewees (nine) had only GCSE or equivalent natural science qualifications; and only two had postgraduate or postdoctoral experience."[131]

---

128 Finamore, personal communication, March 13, 2017.

129. Lausanne Movement, *Cape Town Commitment*, IIA, §6, "Truth, science and emerging technologies." Also, the recommendations of the Presbyterian Church (USA), Advisory Committee on Social Witness Policy, *Power to Change*, 1, urged individuals and families to "Study energy sources, their advantages and disadvantages, and the impacts they have on human communities, all species, and the ecological systems that support life on Earth."

130. Lausanne Movement, *Cape Town Commitment*, IIA, §6 (C).

131. Bouveng and Wilkinson, "Going Beyond the How and Why," 102. Rebecca

They do not claim, however, that this sample is representative of senior leaders across England. Nevertheless, their research is interesting.

And what will churches do in preparation for the new nuclear build programs when individual workers—perhaps hundreds—arrive in their area to construct new nuclear plants?

Local Support for Nuclear Workers

A couple of years ago I was talking with a Baptist church leader whom I knew from my days working at Bristol Baptist College. I shared my experience of working in the nuclear industry and especially about arriving as a young scientist at Hinkley Point B during the time the plant was being commissioned. He shared his experience of how he was involved in working with churches in preparation for the workers that were expected to arrive in the area. It was a valuable exchange.

After I started at Hinkley Point B I joined a small local church where I was warmly welcomed. Here I was given opportunities to preach and later I was invited to join the local Methodist Circuit to preach and lead meetings. After a while I became interested in learning New Testament Greek and started a distance-learning course. This eventually led on to me studying part-time for my first theology degree (although by then I had relocated for my job).

But why am I mentioning my experience all those decades ago? Well, let us think how we will, or should, respond to nuclear workers. Will individuals and churches be hospitable or hostile? Will they welcome the workers into their Christian communities as disciples and worshipers, or wish they had walked past and attended another church? Will church communities be open enough to learn from them about nuclear technologies as well as their individual lives (as the Lausanne statement states—see above)?

Okay, some people will not know any nuclear workers, but will want to know more about nuclear energy, so as to be more engaged in the national and international dialogue on low-carbon energy. What can they do (besides reading my book!)? How about going to the local visitor center of a nuclear plant? In the UK, EDF Energy has eight interactive visitor centers for the general public and school visitors.[132] Entry is free of charge. By prior

---

Bouveng is Assistant Principal at Ustinov College, Durham University, UK, and David Wilkinson is Principal of St. John's College and Professor in the Department of Theology and Religion at Durham University. GCSEs are awards for school children aged 16.

132. EDF Energy, "Visitor Centres."

arrangement it is also possible to visit one of their nuclear plants. How about arranging that with a church group? How about inviting a nuclear industry speaker to a church group?

Could we pray for the new and current workforce? Can we forgive the "nuclear industry" for its perceived past failures? After all, the church has had its own share of unpleasant failures in the past and present.

Hope

Mark Lynas noted: "My conclusion in this book [*Nuclear 2.0*] is ultimately an optimistic one. This is not another lamentation of doom. With an Apollo Program scale-up of nuclear and other low-carbon power sources we still—just about—have time to avoid the worst of global warming. But this will require a lot of burying hatchets."[133]

Christians have written and spoken on hope in the face of possible catastrophic climate change and how we can combat climate change.[134] Recently, John Weaver encouraged readers when he wrote in the face of climate change: "There is hope. Our hope is focused and centred on God," and "Christians are called to a hopeful discipleship in the light of our ultimate hope in God's promises and purposes."[135] We need to remember this in difficult times.

Conclusion

Nuclear alone will not save us from climate change. All technologies for generating electricity have their pros and cons. And electricity itself is dangerous if not used wisely. But with renewables alongside as a companion, nuclear energy can considerably reduce greenhouse gas emissions and air pollution, while providing secure electricity to improve people's lives. In contrast to negative nuclear narratives that I have read, my hope and prayer is that my more positive nuclear narrative will help you to develop your thinking and fairly assess the benefits of nuclear as part of an energy mix.

---

133. Lynas, *Nuclear 2.0* (2013), 9.

134. For example, see the 2013 articles: Margot Hodson, "Editorial: Discovering a Robust Hope"; Martin Hodson, "Losing Hope?" (Margot and Martin are married); and Weaver, "Exploring Hope."

135. Weaver, "Urgent: Christians and Climate Change" (2016). John Weaver is a Baptist minister, former Principal of South Wales Baptist College, and also Dean of the Faculty of Theology at Cardiff University. He is currently Chair of the John Ray Initiative.

LET THERE BE LIGHT!

Further Reading

Berkman, Mark, and Dean Murphy. *Pennsylvania Nuclear Power Plants' Contribution to the State Economy*. Brattle Group, December 2016. http://www.pachamber.org/assets/pdf/advocacy/2016_12_05_pa_nuclear_report.pdf.

EDF Energy. "Hinkley Point C: How the Facts Stack Up." 2016. https://www.edfenergy.com/energy/nuclear-new-build-projects/hinkley-point-c/facts.

Global Nexus Initiative. *Nuclear Power for the Next Generation: Addressing Energy, Climate, and Security Challenges*. Full Report. April 2017 NEI and Partnership for Global Security. http://globalnexusinitiative.org/wp-content/uploads/2017/04/GNI-Digital-Report-single.pdf. (For the Executive Summary see http://globalnexusinitiative.org/wp-content/uploads/2017/04/GNI_Executive-Summary.pdf.)

International Atomic Energy Agency. *Atoms for Peace and Development: How the IAEA Supports the Sustainable Development Goals*. 2015. https://www.iaea.org/sites/default/files/sdg-brochure_forweb.pdf.

Korsnick, Maria. *Nuclear Power Is Critical Infrastructure, Maria Korsnick, Feb. 9, 2017*. 2017 Wall Street Briefing by Maria Korsnick President and Chief Executive Officer NEI February 9, 2017. https://www.nei.org/News-Media/Speeches/Nuclear-Power-is-Critical-Infrastructure. (Briefing given for the NEI "Annual Update on Nuclear Energy in America" and is at: https://www.youtube.com/watch?v=3LOzmdENAns.)

MacKay, David J. C. "David MacKay—Last Interview and Tribute." Video of interview by Mark Lynas, April 3, 2016. Posted April 27, 2016. http://www.marklynas.org/2016/04/david-mackay-last-interview-tribute/. David MacKay wrote *Sustainable Energy—Without the Hot Air*. Shortly before his death he gave his views on the UK's electricity options and showed support for nuclear energy.

Nuclear Energy Agency. *2016 NEA Annual Report*. NEA No. 7349. OECD, 2016. https://www.oecd-nea.org/pub/activities/ar2016/ar2016.pdf. (This has useful articles, e.g. "The Economic Challenges of Nuclear Energy.")

Nuclear Industry Association. *Plutonium Management*. Briefing paper. November 2015. https://www.niauk.org/wp-content/uploads/2016/09/Plutonium-Management.pdf.

Nuclear Institute. *Nuclear Institute Response to the House of Lords Inquiry into Priorities for Nuclear Research and Technologies*. February 22, 2017. http://www.nuclearinst.com/write/MediaUploads/PDFs/NI_-_Response_to_House_of_Lords_Inquiry.pdf. (This includes discussion on SMRs.)

Questions

1. Check out local nuclear power plants in your country and their status (e.g., operational or planned). Use the interactive "Nuclearplanet" web page (http://www.nuklearforum.ch/de/en/nuclearplanet). Alternatively, for the US, see NEI, "Map of US Nuclear Plants." How many nuclear plants are located near to where you live?

2. What is the electrical output from the operational plants?

3. What will be the electrical output from any planned plants?
4. How many nuclear plants are permanently closed down?
5. Determine what types of reactors they are and identify any issues that interest you.
6. Watch the presentation by Maria Korsnick (see Further Reading above) on *Nuclear Power Is Critical Infrastructure* (https://www.youtube.com/watch?v=3LOzmdENAns). In what ways does her presentation help you to understand the benefits of nuclear energy in generating electricity?
7. "The American Nuclear Society (ANS) believes that the sustained operation of the current nuclear fleet is vital to the continued security and economic prosperity of the United States and the world. U.S. nuclear power plants provide reliable clean energy, help diversify our electricity supply, and support continued U.S. influence over global safety and nonproliferation standards." (ANS, *U.S. Commercial Nuclear Power Plants: A Vital National Asset*. Position Statement #26, April 2017. http://cdn.ans.org/pi/ps/docs/ps26.pdf.) Read the short ANS position statement and reflect on their reasons, recommendations, and resulting benefits for the U.S. and the world.
8. In Luke 10:25–37 Jesus talked with a lawyer on the the requirement to "Love your neighbor as yourself." Then to answer the question "And who is my neighbor?," Jesus told "The Parable of the Good Samaritan," about how people should act. Suggest suitable actions, or strategies, for Christians, Churches, and Church leaders to display towards people in the nuclear energy industry who generate clean, low-carbon electricity on our behalf. Can we employ a more positive narrative (a new nuclear narrative) instead of a negative nuclear narrative?

Glossary

FOR MORE INFORMATION, CONSULT the *Glossary of Nuclear Terms*, edited by Gareth Davies (London/Bristol: Burges Salmon, June 2015; https://www.burges-salmon.com/-/media/files/publications/open-access/burges_salmon_glossary_of_nuclear_terms_june_2015.pdf). For basic information on some reactors, with schematic diagrams, consult the World Nuclear Association, *Nuclear Power Reactor Characteristics. 2016/17 Pocket Guide. August 2016.* http://www.world-nuclear.org/getmedia/80f869be-32c8-46e7-802d-eb4452939ec5/Pocket-Guide-Reactors.pdf.aspx.

Word	Explanation
absorbed dose	The energy deposited in any medium by any ionizing radiation. Expressed in units called grays (Gy)—see below.
alpha particle	Made up of 2 neutrons and 2 protons (i.e., the helium-4 nucleus). Inside the human body alpha particles are damaging to tissue.
atom	Smallest part of an element that still retains its properties. Made up of the nucleus and orbital electrons. Normally, the number of electrons is the same as the number of protons, so the atom is electrically neutral.
beta particle	An electron that originates from the nucleus during radioactive decay, when a neutron changes into a proton and emits the electron.
becquerel	SI unit of radioactivity (Bq) of one nuclear disintegration per second.
collective dose	The sum of individual doses in a population.
curie	Older unit of radioactivity (Ci) equal to 3.7×10^{10} becquerels (Bq).

GLOSSARY

Word	Explanation
dose	In radiation protection, the term dose needs explanation. It is not the same, for example, as taking a dose of medicine. Dose is expressed as absorbed dose, in grays, where 1 Gy = 1 joule of energy deposited per kilogram. But more frequently in radiation protection the unit sievert (Sv) is used (see below).
effective dose	Effective dose is obtained by multiplying the equivalent dose (in sieverts) by organ/tissue weighting factors and summing them. For example, with radon exposure of the lung the dose to the lung is multiplied by the weighting factor for the lung.
electron	Electrons are negatively charged and orbit the central nucleus of the atom in defined shells. They are much lighter than neutrons and protons, being just 1/1,840 of their mass.
fission	The process by which a heavy nucleus is split into normally two lighter elements. Energy is released as well as neutrons (which may go on to cause more fissions and in certain circumstances cause a chain reaction).
fusion	The process by which lighter elements are fused together to give heavier elements. Energy is released. Our sun works by nuclear fusion.
gamma radiation	Electromagnetic radiation (like light) that originates in the nucleus of atoms. Gamma rays are penetrating and can go through the human body.
neutron	Neutrons are electrically neutral with about the same mass as the proton. They reside in the nucleus. Apart from hydrogen-1 (which only has one proton in the nucleus) elements have neutrons in their nucleus.
proton	Protons are positively charged with about the same mass as the neutron. They reside in the nucleus.
rad	Older unit of absorbed dose, where 1 rad = 0.01 joules per kilogram.
radiation	I use the term radiation as a shorthand for ionizing radiation. This radiation interacts with orbital electrons in the atoms and causes changes which can affect cells.
rem	Older unit of radiation dose, where 1 rem = 10 mSv.
sievert	SI unit of radiation dose (Sv). Sieverts are obtained by multiplying the absorbed dose (in grays) by radiation weighting factors. These take account of the fact that some radiation releases its energy to orbital electrons in short dense tracks. For example, inside the body alpha particles are more damaging than beta particles. The term is called equivalent dose, also measured in sieverts (Sv).
x-rays	Electromagnetic radiation (like light) that originates from orbital electrons. X-rays are penetrating and can go through the human body.

Bibliography

Abate, Tom. "Stanford Researchers Say Extracting Uranium from Seawater Could Help Nuclear Power Play a Larger Role in a Carbon-Free Energy Future." *Stanford News*, February 17, 2017. http://news.stanford.edu/2017/02/17/uranium-seawater-factors-nuclear-power/.

Adachi, N., et al. "Measurement and Comparison of Individual and External Doses of High-School Students in Japan, France, Poland and Belarus—the 'D-Shuttle' Project–." *Journal of Radiological Protection* 36.1 (2016) 49–66. http://iopscience.iop.org/article/10.1088/0952-4746/36/1/49.

Adams, Jarret. "What's Really Killing America's Nuclear Plants." *World Nuclear News*, January 30, 2017. http://www.world-nuclear-news.org/V-Whats-really-killing-Americas-nuclear-plants-30011701.html.

Adams, Rod. "Continuing Education on Advanced Reactors for U.S. Senate." Atomic Insights, May 18, 2016. http://atomicinsights.com/continuing-education-advanced-reactors-u-s-senate/.

Adams, Steve, and Jonathan Allday. *Advanced Physics*. Oxford: Oxford University Press, 2000.

Advisory Committee on Social Witness Policy. *The Power to Change: U.S. Energy Policy and Global Warming*. Office of the General Assembly, Presbyterian Church (USA), 2008. http://www.pcusa.org/media/uploads/acswp/pdf/energyreport.pdf.

AIP (American Institute of Physics). "Radiation Exposure of U.S. Population: American Association of Physicists in Medicine Comments on Controversial Report." *ScienceDaily*, March 5, 2009. https://www.sciencedaily.com/releases/2009/03/090303123812.htm.

Alexander, Denis. *Rebuilding the Matrix: Science and Faith in the 21st Century*. Oxford: Lion, 2001.

Allison, Wade. *Nuclear Is for Life: A Cultural Revolution*. N.p: Wade Allison Publishing, 2015. http://www.nuclear4life.com/first.

Alvarez, Luis E., et al. "Radiation Dose to the Global Flying Population." *Journal of Radiological Protection* 36.1 (2016) 93–103. http://iopscience.iop.org/article/10.1088/0952-4746/36/1/93.

BIBLIOGRAPHY

American Lung Association, et al. *The National Radon Action Plan: A Strategy for Saving Lives.* EPA (US), November 2015. https://www.epa.gov/sites/production/files/2015-11/documents/nrap_guide_2015_final.pdf.

Anglican Consultative Council, and Anglican Communion Environmental Network. *The World Is Our Host: A Call to Urgent Action for Climate Justice.* Volmoed Conference and Retreat Centre, South Africa, February 23–27, 2015. http://acen.anglicancommunion.org/media/148818/The-World-is-our-Host-FINAL-TEXT.pdf.

Anglican Communion. "Marks of Mission." http://www.anglicancommunion.org/identity/marks-of-mission.aspx.

ANS (American Nuclear Society). "Oklo's Natural Fission Reactors." July 12, 2012. http://www.ans.org/pi/np/oklo/#Q1.

———. "Radiation Dose Calculator." Uses earlier units (millirems). February 8, 2016. http://www.ans.org/pi/resources/dosechart/.

Archer, Louise, et al. *ASPIRES: Young People's Science and Career Aspirations, Age 10–14.* ASPIRES Project, November 2013. http://www.kcl.ac.uk/sspp/departments/education/research/aspires/ASPIRES-final-report-December-2013.pdf.

———. *Ten Science Facts & Fictions: The Case for Early Education about STEM Careers.* ASPIRES Project, 2012. http://www.kcl.ac.uk/sspp/departments/education/research/aspires/10FactsandFictionsfinalversion.pdf.

Asafu-Adjaye, John, et al. *An Ecomodernist Manifesto.* April 2015. http://www.ecomodernism.org/manifesto-english/.

Atkinson, David. *Climate Change and the Gospel: Why We in the Churches Need to Treat Climate Change More Urgently.* Operation Noah, January 6, 2015. http://operationnoah.org/wp-content/uploads/2015/02/Climate-and-Gospel-David-Atkinson-30-01-2015.pdf.

———. "Well-Meaning but Somewhat Naïve"? A Response to the GWPF Briefing Paper: *The Papal Encyclical—A Critical Christian Response by Bernard Donoughue and Peter Forster.* Operation Noah, August 20, 2015. http://operationnoah.org/wp-content/uploads/2015/08/Atkinson-response-to-GWPF-paper-20-21-08-2015.pdf.

Ausubel, J. H. *Power Density and the Nuclear Opportunity.* Revised December 7, 2015. https://phe.rockefeller.edu/docs/PowerDensity_Final120815.pdf.

———. "Renewable and Nuclear Heresies." *International Journal of Nuclear Governance, Economy and Ecology* 1.3 (2007) 229–43.

Baker, John Austin, et al. *The Church and the Bomb: Nuclear Weapons and Christian Conscience.* Report to the Church of England's Board for Social Responsibility. London: Hodder & Stoughton, CIO, 1982.

Balonov, Mikhail. "The Chernobyl Accident as a Source of New Radiological Knowledge: Implications for Fukushima Rehabilitation and Research Programmes." *Journal of Radiological Protection* 33.1 (2013) 27–40. http://iopscience.iop.org/article/10.1088/0952-4746/33/1/27.

———. "On Protecting the Inexperienced Reader from Chernobyl Myths." *Journal of Radiological Protection* 32.2 (2012) 181–89. http://iopscience.iop.org/article/10.1088/0952-4746/32/2/181.

Balter, Stephen, et al. "Radiation Is Not the Only Risk." *American Journal of Roentgenology* 196.4 (April 2011) 762–67. http://www.ajronline.org/doi/pdf/10.2214/AJR.10.5982.

Bartlett, D. T., et al. *The Health Protection Agency Radiation Protection Division Passive Survey Instrument.* HPA-RPD-015. HPA, August 2006. https://www.gov.uk/government/uploads/system/uploads/attachment_data/file/340196/HpaRpd015.pdf.

BIBLIOGRAPHY

Bauckham, Richard. *Bible and Ecology: Rediscovering the Community of Creation.* Sarum Theological Lectures. London: Darton, Longman and Todd, 2010.

Bawden, Tom. "How America Uses Russian Warheads to Heat Its Homes." *Independent*, December 27, 2013. http://www.independent.co.uk/news/world/americas/how-america-uses-russian-warheads-to-heat-its-homes-9027617.html.

Baylon, Caroline, et al. *Cyber Security at Civil Nuclear Facilities: Understanding the Risks.* The Royal Institute of International Affairs, Chatham House, September 2015. https://www.chathamhouse.org/sites/files/chathamhouse/field/field_document/20151005CyberSecurityNuclearBaylonBruntLivingstoneUpdate.pdf. See also the Executive Summary, available at https://www.chathamhouse.org/publication/cyber-security-civil-nuclear-facilities-understanding-risks#.

BBC (British Broadcasting Corporation). "1966: Aberfan—A Generation Wiped Out." On This Day. http://news.bbc.co.uk/onthisday/hi/witness/october/21/newsid_3194000/3194860.stm.

BEIS (Department for Business, Energy and Industrial Strategy). *Building our Industrial Strategy.* Green paper, January 2017. https://beisgovuk.citizenspace.com/strategy/industrial-strategy/supporting_documents/buildingourindustrialstrategygreenpaper.pdf.

———. *Energy and Climate Change Public Attitudes Tracker: Wave 18.* Summary of key findings. July 2016. https://www.gov.uk/government/collections/public-attitudes-tracking-survey.

———. *Energy and Climate Change Public Attitude Tracker: Wave 19.* October 2016. https://www.gov.uk/government/uploads/system/uploads/attachment_data/file/563236/Summary_of_key_findings_BEIS_Public_Attitudes_Tracker_-_wave_19.pdf.

———. *Energy and Climate Change Public Attitude Tracker: Wave 20.* February 2017. https://www.gov.uk/government/uploads/system/uploads/attachment_data/file/590505/Summary_of_key_findings_BEIS_Public_Attitudes_Tracker_-_wave_20.pdf.

———. *The Energy Revolution and Future Challenges for UK Energy and Climate Change Policy: Government Response to the Energy and Climate Change Committee's Third Report of Session 2016–17: Fourth Special Report—Appendix: Government Response.* December 2016. https://www.publications.parliament.uk/pa/cm201617/cmselect/cmbeis/945/94502.htm.

———. *Energy Trends March 2017.* https://www.gov.uk/government/uploads/system/uploads/attachment_data/file/604690/Energy_Trends_March_2017.pdf.

———. *UK Energy in Brief 2016.* July 2016. https://www.gov.uk/government/uploads/system/uploads/attachment_data/file/540135/UK_Energy_in_Brief_2016_FINAL.pdf.

———. *UK Energy Statistics—2016 Provisional Data.* Statistical Press Release, February 23, 2017. https://www.gov.uk/government/uploads/system/uploads/attachment_data/file/593413/Press_Notice_February_2017.pdf.

Berkman, Mark, and Dean Murphy. *The Nuclear Industry's Contribution to the U.S. Economy.* Brattle Group. July 7, 2015. http://www.brattle.com/system/news/pdfs/000/000/895/original/The_Nuclear_Industry%27s_Contribution_to_the_U.S._Economy.pdf?1436280444.

BIBLIOGRAPHY

———. *Pennsylvania Nuclear Power Plants' Contribution to the State Economy.* Brattle Group, December 2016. http://www.pachamber.org/assets/pdf/advocacy/2016_12_05_pa_nuclear_report.pdf.

Billiere, Edward de la. "Going Nuclear in South Australia?" *World Nuclear News*, November 18, 2016. http://www.world-nuclear-news.org/V-Going-nuclear-in-South-Australia-18111601.html.

Birmingham Policy Commission. *The Future of Nuclear Energy in the UK.* University of Birmingham, July 2012. http://www.birmingham.ac.uk/Documents/research/SocialSciences/NuclearEnergyFullReport.pdf.

BIS (Department for Business, Innovation and Skills). "Government Confirms £80 Million for National Colleges to Deliver the Workforce of Tomorrow." May 9, 2016. https://www.gov.uk/government/news/government-confirms-80-million-for-national-colleges-to-deliver-the-workforce-of-tomorrow.

———. *Nuclear Industrial Strategy: The UK's Nuclear Future.* BS/13/627. March 2013. https://www.gov.uk/government/uploads/system/uploads/attachment_data/file/168048/bis-13-627-nuclear-industrial-strategy-the-uks-nuclear-future.pdf.

BIS and DECC (Department for Business, Innovation and Skills; and Department of Energy and Climate Change). *Nuclear Energy Research and Development Roadmap: Future Pathways.* 2013. https://www.gov.uk/government/publications/nuclear-energy-research-and-development-roadmap-future-pathways.

Bisconti Research. *6th Biennial Survey of U.S. Nuclear Power Plant Neighbors: Broad and Deep Support for Nuclear Energy and the Nearby Plant Continues.* Summer 2015. http://www.nei.org/CorporateSite/media/filefolder/Backgrounders/Reports-Studies/2015-Plant-Neighbor-Memo-070715.pdf.

Bookless, Dave. *Planetwise: Dare to Care for God's World.* Nottingham: Inter-Varsity, 2008.

Bosch, David J. *Transforming Mission: Paradigm Shifts in Theology of Mission.* American Society of Missiology Series 16. New York: Orbis, 1991.

Boston, Andy. "Can Science Save the World?" *Mission Catalyst* 3 (2014) 12–13. http://www.bmsworldmission.org/engagecatalyst/mission-catalyst-climate-change-is-boring/can-science-save-the-world.

Bouveng, Rebecca, and David Wilkinson. "Going Beyond the How and Why of Science-Religion? Senior Christian Leaders on Science and Personal Faith." *Science and Christian Belief* 28.2 (2016) 100–16. https://www.scienceandchristianbelief.org/view_abstract.php?ID=1316.

Boyle, Godfrey, et al., eds. *Energy Systems and Sustainability.* Oxford: Oxford University Press and the Open University, 2003.

Brand, Stewart. *Whole Earth Discipline: Why Dense Cities, Nuclear Power, Genetically Modified Crops, Restored Wildlands, Radical Science, and Geoengineering Are Essential.* Rev. ed. London: Atlantic, 2010.

Bridgwater and Taunton College. "National College for Nuclear Announced." March 20, 2015. http://www.bridgwater.ac.uk/news-article.php?id=2113.

Bromet, Evelyn J. "Mental Health Consequences of the Chernobyl Disaster." *Journal of Radiological Protection* 32.1 (2012) N71–N75. http://iopscience.iop.org/article/10.1088/0952-4746/32/1/N71.

Brook, Barry W. "Could Nuclear Fission Energy, Etc., Solve the Greenhouse Problem? The Affirmative Case." *Energy Policy* 42 (2012) 4–8. http://350.me.uk/TR/Hansen/BarryBrook.pdf.

BIBLIOGRAPHY

Brown, Ed. *Climate Change after Paris: What It Means for the Evangelical Church*. Lausanne Global Analysis 5.3 (May 2016). https://www.lausanne.org/content/lga/2016-05/climate-change-after-paris.

Brown, Lester R., et al. *The Great Transition: Shifting from Fossil Fuels to Solar and Wind Energy*. New York: Norton, 2015.

Brown, M. J. "Too Cheap to Meter?" Canadian Nuclear Society, February 20, 2009. https://cns-snc.ca/media/media/toocheap/toocheap.html.

Cabinet Office. *HMG Security Policy Framework*. April 2014. https://www.gov.uk/government/uploads/system/uploads/attachment_data/file/316182/Security_Policy_Framework_-_web_-_April_2014.pdf.

———. *Prospectus: Introducing the National Cyber Security Centre*. Government Communications Headquarters and CESG, May 2016. https://www.gov.uk/government/uploads/system/uploads/attachment_data/file/525410/ncsc_prospectus_final_version_1_0.pdf.

———. *The UK Cyber Security Strategy 2011–2016: Annual Report*. April 2016. https://www.gov.uk/government/uploads/system/uploads/attachment_data/file/516331/UK_Cyber_Security_Strategy_Annual_Report_2016.pdf.

Caplin, Mark. "Bishop of Hereford Speaks Up for Nuclear Power." *Christianity Today*, April 25, 2013. http://www.christiantoday.com/article/bishop.of.hereford.speaks.up.for.nuclear.power/32235.htm.

Cartlidge, Edwin. "US Set for Borehole Field Trial." *Physics World*, April 4, 2016. http://live.iop-pp01.agh.sleek.net/2016/03/18/us-set-for-borehole-field-trial/.

Cavnar, Bob. *Disaster on the Horizon: High Stakes, High Risks, and the Story Behind the Deepwater Well Blowout*. White River Junction, VT: Chelsea Green, 2010.

CCC (Committee on Climate Change). *Power Sector Scenarios for the Fifth Carbon Budget*. October 2015. https://d2kjx2p8nxa8ft.cloudfront.net/wp-content/uploads/2015/10/Power-sector-scenarios-for-the-fifth-carbon-budget.pdf.

Cefas (Centre for Environment, Fisheries and Aquaculture Science). *Radioactivity in Food and the Environment, 2014* (RIFE – 20). Compiled for the Environment Agency et al. October 2015. http://www.food.gov.uk/sites/default/files/rife-20.pdf.

———. *Radioactivity in Food and the Environment, 2015* (RIFE – 21). Compiled for the Environment Agency et al. October 2016. https://www.food.gov.uk/sites/default/files/rife-2015.pdf.

Celebi, Martin, et al. *Nuclear Retirement Effects on CO_2 Emissions: Preserving a Critical Clean Resource*. Brattle Group, December 2016. http://brattle.com/system/news/pdfs/000/001/158/original/Brattle_Nuclear-Carbon_Whitepaper_-_Dec2016.pdf.

Chatzis, Irene. "Decommissioning and Environmental Remediation: An Overview." *IAEA Bulletin* 57.1 (April 2016) 4–5. https://www.iaea.org/publications/magazines/bulletin/57-1.

Chen, Jing, and Deborah Moir. "An Estimation of the Annual Effective Dose to the Canadian Population from Medical CT Examinations." *Journal of Radiological Protection* 30.2 (2010) 131–37. http://iopscience.iop.org/article/10.1088/0952-4746/30/2/002.

Cheng, Huiping, et al. "HPR1000 – Advanced Pressurised Water Reactor with Active and Passive Safety." *Nuclear Future* 12.4 (2016) 38–43.

Chow Y., et al. *Evaluation and Equity Audit of the Domestic Radon Programme in England*. HPA-CRCE-013. HPA, February 2011. https://www.gov.uk/government/uploads/system/uploads/attachment_data/file/340156/HPA-CRCE-013_for_website.pdf.

BIBLIOGRAPHY

Christian Aid. *The Big Shift Q&A.* August 2015. http://www.christianaid.org.uk/Images/The-Big-Shift-QandA-august-2015.pdf.

Christianto, Victor, and Florentin Smarandache. *A Christian Ethics Consideration on Nuclear Energy.* 2nd rev. ResearchGate, July 2, 2013. https://www.researchgate.net/publication/244478561_A_Christian_Ethics_consideration_on_Nuclear_Energy.

Church of England. "Archbishop of Canterbury Join Faith Leaders in Call for Urgent Action to tackle Climate Change." June 16, 2015. https://www.churchofengland.org/media-centre/news/2015/06/archbishop-of-canterbury-join-faith-leaders-in-call-for-urgent-action-to-tackle-climate-change.aspx.

———. "Church of England Welcomes Climate Encyclical." June 18, 2015. https://www.churchofengland.org/media-centre/news/2015/06/church-of-england-welcomes-climate-encyclical.aspx.

———. *Report of Proceedings 2015: General Synod July Group of Sessions.* 46.2. https://www.churchofengland.org/media/2445222/rop_july_2015__final_indexed_version_.pdf.

Church of Scotland Church and Society Council. *Energy for a Changing Climate.* Report presented to the General Assembly, May 2007. http://www.srtp.org.uk/assets/uploads/Energy_for_a_Changing_Climate_2007.pdf.

Clegg, Richard, and Mamdouh El-Shanawany. "Small Modular Reactors." *Ingenia* 52 (September 2012) 46–50. http://www.ingenia.org.uk/Ingenia/Articles/794.

Climate Stewards. *London–New York Return.* http://www.climatestewards.org/downloads/cs_flier_flying_front.jpg.

CMI (Chartered Management Institute). *Managers and the Moral Maze: A Summary of Findings from a Survey on Workplace Ethics.* September 2013. http://www.managers.org.uk/~/media/Research%20Report%20Downloads/Managers_and_the_moral_maze_Sept13.pdf.

COC (Committee on Carcinogenicity of Chemicals in Food, Consumer Products and the Environment). *Statement on Consumption of Alcoholic Beverages and Risk of Cancer.* Statement 2015/S2. PHE, Centre for Radiation, Chemical and Environmental Hazards, 2015. https://www.gov.uk/government/uploads/system/uploads/attachment_data/file/490584/COC_2015_S2__Alcohol_and_Cancer_statement_Final_version.pdf.

Cole, P., et al. "Strategies for Engaging with Future Radiation Protection Professionals: A Public Outreach Case Study." *Journal of Radiological Protection* 35.4 (2015) N25–N32. http://iopscience.iop.org/article/10.1088/0952-4746/35/4/N25.

Connor, Steve. "Nuclear Power? Yes Please..." *Independent,* February 22, 2009. http://www.independent.co.uk/environment/green-living/nuclear-power-yes-please-1629327.html.

Cooper, Tim. *Faith and Power: The Case for a Low Consumption, Non-Nuclear, Energy Strategy.* Christian Ecology Link, March 2006. http://www.greenchristian.org.uk/wp-content/uploads/2013/02/fp-2006-2013.pdf.

———. *Green Christianity: Caring for the Whole Creation.* London: Spire, 1990.

Corner, Adam, et al. "Nuclear Power, Climate Change and Energy Security: Exploring British Public Attitudes." *Energy Policy* 39 (2011) 4823–33. http://www.svsu.edu/~gmlange/BJLG09F11.pdf.

Cornwall Alliance. "An Evangelical Declaration on Global Warming." May 1, 2009. http://cornwallalliance.org/2009/05/evangelical-declaration-on-global-warming/.

———. "An Open Letter to Pope Francis on Climate Change." April 27, 2015. http://cornwallalliance.org/anopenlettertopopefrancisonclimatechange/.

BIBLIOGRAPHY

———. "An Open Letter to the Signers of 'Climate Change: An Evangelical Call To Action' and Others Concerned about Global Warming." July 25, 2006. http://cornwallalliance.org/2006/07/an-open-letter-to-the-signers-of-climate-change-an-evangelical-call-to-action-and-others-concerned-about-global-warming-2/.

Cox, Brian, and Andrew Cohen. *Human Universe*. London: William Collins, 2014.

———. *Wonders of the Universe*. London: HarperCollins, 2011.

CPNI (Centre for the Protection of National Infrastructure). *Personal Security: An Ongoing Responsibilty*. 2015. https://www.cpni.gov.uk/system/files/documents/5b/04/ongoing-personnel-security-infographic.pdf.

Cravens, Gwyneth. *Power to Save the World: The Truth about Nuclear Energy*. New York: Vintage, 2008. http://cravenspowertosavetheworld.com/.

Dauer, Lawrence T., et al. "Fears, Feelings, and Facts: Interactively Communicating Benefits and Risks of Medical Radiation With Patients." *American Journal of Roentgenology* 196.4 (2011) 756–71. http://www.ajronline.org/doi/full/10.2214/AJR.10.5956.

Davies, Gareth, ed. *Glossary of Nuclear Terms*. London/Bristol: Burges Salmon, June 2015. https://www.burges-salmon.com/-/media/files/publications/open-access/burges_salmon_glossary_of_nuclear_terms_june_2015.pdf.

Davies, Paul. *The Goldilocks Enigma: Why Is the Universe Just Right for Life?* London: Allen Lane, 2006.

Davis, E. D., et al. "Oklo Reactors and Implications for Nuclear Science." *International Journal of Modern Physics E* 23.4 1430007 (2014). http://arxiv.org/abs/1404.4948.

de-la-Noy, Michael. "The Rt Rev Hugh Montefiore." *Guardian*, May 14, 2005. http://www.theguardian.com/news/2005/may/14/guardianobituaries.religion.

DECC (Department of Energy and Climate Change). *The Justification of Practices Involving Ionising Radiation Regulations 2004*. URN 10D/830. October 2010. https://www.gov.uk/government/uploads/system/uploads/attachment_data/file/47935/667-decision-ap1000-nuclear-reactor.pdf.

———. *National Policy Statement for Nuclear Power Generation (EN-6)*. Vol. 1 of 2. URN 11D/716. London: TSO, 2011. https://www.gov.uk/government/uploads/system/uploads/attachment_data/file/47859/2009-nps-for-nuclear-volumeI.pdf.

———. "New Nuclear Will Learn Lessons from Past Mistakes—Davey." March 2, 2012. https://www.gov.uk/government/news/new-nuclear-will-learn-lessons-from-past-mistakes-davey.

———. *Overarching National Policy Statement for Energy (EN-1)*. URN 11D/711. London: TSO, 2011. https://www.gov.uk/government/uploads/system/uploads/attachment_data/file/47854/1938-overarching-nps-for-energy-en1.pdf.

———. *Small Modular Reactors. Competition: Phase One*. URN 16D/045. March 17, 2016. https://www.gov.uk/government/uploads/system/uploads/attachment_data/file/508616/SMR_Competition_Phase_1_Guidance.pdf.

DECC (Department of Energy and Climate Change), et al. *Strategy for the Management of Naturally Occurring Radioactive Material (NORM) Waste in the United Kingdom*. Edinburgh: APS Group Scotland for the Scottish Government, July 2014. https://www.gov.uk/government/uploads/system/uploads/attachment_data/file/335821/Final_strategy_NORM.pdf.

———. *Strategy for the Management of Solid Low Level Radioactive Waste from the Non-Nuclear Industry in the United Kingdom. Part 1 – Anthropogenic Radionuclides*. URN 12D/026. DECC, 2011. https://www.gov.uk/government/uploads/system/uploads/attachment_data/file/48291/4616-strategy-low-level-radioactive-waste.pdf.

Deepwater Horizon Study Group. *Final Report on the Investigation of the Macondo Well Blowout*. Center for Catastrophic Risk Management, University of California Berkeley March 1, 2011. http://ccrm.berkeley.edu/pdfs_papers/bea_pdfs/DHSGFinalReport-March2011-tag.pdf.

Deis, Robert. "'Too Cheap to Meter'—The Infamous Nuclear Power Misquote . . .'" September 17, 2015. http://www.thisdayinquotes.com/2009/09/too-cheap-to-meter-nuclear-quote-debate.html.

Deju, Raul A., et al. *Nuclear Is Hot!: Everything You Wanted to Know about Nuclear Science but Were Afraid to Ask*. N.p.: Energy Solutions Foundation, 2009.

Department for Culture, Media and Sport. "Two Thirds of Large UK Businesses Hit by Cyber Breach or Attack in Past Year." May 8, 2016. https://www.gov.uk/government/news/two-thirds-of-large-uk-businesses-hit-by-cyber-breach-or-attack-in-past-year.

Dobson, Andrew. *Environmental Politics: A Very Short Introduction*. Oxford: Oxford University Press, 2016.

Done, Chris. *Thinking About . . . the Big Bang: How Did the Universe Begin?* Thinking About . . . leaflet series. Christians in Science, 2016. http://www.cis.org.uk/resources/thinking/.

Drijfhout, Sybren, et al. "Catalogue of Abrupt Shifts in Intergovernmental Panel on Climate Change Climate Models." *Proceedings of the National Academy of Sciences* 112.43 (2015) 1–10. https://www.researchgate.net/publication/282800885_Catalogue_of_abrupt_shifts_in_Intergovernmental_Panel_on_Climate_Change_climate_models.

DTI (Department of Trade and Industry). *Our Energy Future – Creating a Low Carbon Economy*. Energy white paper 2003. February 2003. http://webarchive.nationalarchives.gov.uk/+/http:/www.berr.gov.uk/energy/whitepaper/2003/page21223.html.

Duckworth, Anna. "Q&A with Anna Duckworth." *Innovate* 9 (May 2016) 3–4. http://www.nnl.co.uk/media/2064/innovate-9-may-2016-final-web-version.pdf.

Dunster, H. J., et al. "District Surveys Following the Windscale Incident, October 1957." *Journal of Radiological Protection* 27.3 (2007) 217–30. http://iopscience.iop.org/article/10.1088/0952-4746/27/3/001.

Durham University. "Individual Module Outlines (by Module Title)." February 27, 2017. https://www.dur.ac.uk/common.awards/modules/outlines.titles/.

———. "Programme Documents: Trinity College with Bristol Baptist College." 2016/17. https://www.dur.ac.uk/common.awards/regulations/trinity/.

Dyer, Gwynne. *Climate Wars: The Fight for Survival as the World Overheats*. Oxford: Oneworld, 2010.

EA (Environment Agency). "The 50 Things that Will Save the Planet. Results from a Poll of 25 Experts." *Your Environment* 17, Extra (November 2007–January 2008). http://image.guardian.co.uk/sys-files/Environment/documents/2007/10/31/50top.pdf.

EA (Environment Agency), et al. *New Nuclear Power Stations: GDA*. Infographic March 31, 2017. https://www.gov.uk/government/uploads/system/uploads/attachment_data/file/605324/GDA.pdf.

Earney, Graham, ed. *The Power of God?: Essays in the Theology Relating to Nuclear Power*. N.p.: South West Region of the Industrial Mission Association and the Social Responsibility Group of the Diocese of Bath and Wells, September 1990.

BIBLIOGRAPHY

EC (European Commission). "Commission Presents Report on Investments in Nuclear Safety." April 5, 2016. http://ec.europa.eu/energy/en/news/commission-presents-nuclear-illustrative-programme.

———. "G20 Energy Ministers Commit to Tackle Together Global Energy and Climate Challenges." July 4, 2016. http://ec.europa.eu/energy/en/news/g20-energy-ministers-commit-tackle-together-global-energy-and-climate-challenges.

———. "Nuclear Energy: Safe Nuclear Power." September 16, 2016. http://ec.europa.eu/energy/en/topics/nuclear-energy.

———. *Nuclear Illustrative Programme*. COM(2016) 177 final. April 4, 2016. http://ec.europa.eu/transparency/regdoc/rep/1/2016/EN/1-2016-177-EN-F1-1.PDF.

EDF Energy. *EDF Energy Nuclear Generation: Our Journey towards ZERO HARM. Summary of Our Nuclear Safety and Waste Policies and Management Systems*. September 2015. https://www.edfenergy.com/sites/default/files/our_journey_towards_zero_harm_2015.pdf.

———. *Hinkley Point B Power Station: March 2016 Monthly Report*. https://www.edfenergy.com/sites/default/files/hpb_community_report_february_2016.pdf.

———. *Hinkley Point C: Building Britain's Low-Carbon Future*. July 2016. https://www.edfenergy.com/sites/default/files/hpc_building_britains_low-carbon_future_-_july_2016.pdf.

———. "Hinkley Point C: How the Facts Stack Up." 2016. https://www.edfenergy.com/energy/nuclear-new-build-projects/hinkley-point-c/facts.

———. "Hinkley Point C: Securing the UK's Energy Future." https://www.edfenergy.com/energy/nuclear-new-build-projects/hinkley-point-c.

———. "Inspiring Young Minds." https://www.edfenergy.com/energy/nuclear-new-build-projects/hinkley-point-c/inspire.

———. "Meet the People Living and Working at Hinkley Point C." Video. September 15, 2016. https://www.youtube.com/watch?v=vpJqW6jJIAg.

———. "Office for Nuclear Regulation Publishes List of UK 'Nuclear Reportable Events.'" February 4, 2016. http://media.edfenergy.com/r/1019/office_for_nuclear_regulation_publishes_list_of_uk.

———. "Reporting of Safety Events." https://www.edfenergy.com/energy/safety-event-reporting.

———. "Virtual Tours: Nuclear Generation and Safety." Video. https://www.edfenergy.com/virtual-tours/nuclear-safety.

———. "Visitor Centres." 2016. https://www.edfenergy.com/energy/education/visitor-centres.

EIA (Energy Information Administration [US DOE]). "Glossary." http://www.eia.gov/tools/glossary/.

———. *International Energy Outlook 2016*. DOE/EIA-0484. May 2016. http://www.eia.gov/outlooks/ieo/pdf/0484(2016).pdf.

Electrical Safety First. "For DIYers." 2016. http://www.electricalsafetyfirst.org.uk/guides-and-advice/for-diyers/.

———. "How Safe Is Your Home?" 2014. http://www.electricalsafetyfirst.org.uk/guides-and-advice/around-the-home/how-safe-is-your-home/.

Elliott, Alex. "Issues in Medical Exposures." *Journal of Radiological Protection* 29.2A (2009) A107–21. http://iopscience.iop.org/article/10.1088/0952-4746/29/2A/S07.

Endicott, Neil. *Thorium-Fuelled Molten Salt Reactors*. Report for the All Party Parliamentary Group on Thorium Energy. Weinberg Foundation, 2013. http://www.

the-weinberg-foundation.org/wp-content/uploads/2013/06/Thorium-Fuelled-Molten-Salt-Reactors-Weinberg-Foundation.pdf.

Engineering Council and Royal Academy of Engineering. *Statement of Ethical Principles for the Engineering Profession*. April 2014. http://www.engc.org.uk/engcdocuments/internet/Website/Statement%20of%20Ethical%20Principles.pdf.

EPA (Ireland) (Environmental Protection Agency [Ireland]) Office of Radiological Protection. *Radon and Your Health*. 2016. http://www.epa.ie/pubs/reports/radiation/RPII_Pamphlet_Radon_Health.pdf.

EPA (US) (Environmental Protection Agency [US]). *Basic Radon Facts*. EPA 402/F-12/005. February 2013. https://www.epa.gov/sites/production/files/2016-08/documents/july_2016_radon_factsheet.pdf.

EPSRC (Engineering and Physical Sciences Research Council). *EPSRC Independent Review of Fission and Fusion*. March 2016. https://www.epsrc.ac.uk/newsevents/pubs/indrevfissionfusion/.

ERDO (European Repository Development Organisation) Working Group. *Shared Solutions for Spent Fuel and Radioactive Wastes: Responding to EC Directive 2011/70/EURATOM*. http://www.erdo-wg.eu/Documents_files/ERDO%20Strategy%20Document.pdf.

Evangelical Environmental Network. "Statement by Evangelical Leaders on Mercury and the Unborn." http://www.creationcare.org/mercury_the_unborn.

Fahlquist, Jessica Nihlén, and Sabine Roeser. "Nuclear Energy, Responsible Risk Communication and Moral Emotions: A Three Level Framework." *Journal of Risk Research* 18.3 (2015) 333–46. http://www.tandfonline.com/doi/full/10.1080/13669877.2014.940594.

Farmelo, Graham. *Churchill's Bomb: A Hidden History of Science, War and Politics*. London: Faber, 2013.

Fawaz-Huber, May. "How the United Kingdom Seeks to Enhance Nuclear Security With the Help of IPPAS." *IAEA Bulletin* 57.4 (December 2016) 12–13. https://www.iaea.org/publications/magazines/bulletin/57-4.

Fischhoff, Baruch, and John David Kadvany. *Risk: A Very Short Introduction*. Oxford: Oxford University Press, 2011.

FOE (Friends of the Earth). *Nuclear Power*. July 2015. https://www.foe.co.uk/sites/default/files/downloads/policy-position-nuclear-power-75191.pdf.

———. *A Plan for Clean British Energy: Powering the UK with Renewables—and Without Nuclear*. September 2012. https://www.foe.co.uk/sites/default/files/downloads/plan_cbe_report.pdf.

FOE Cymru (Friends of the Earth Cymru). *Evidence to the Welsh Affairs Committee: Inquiry into Energy in Wales*. 2006. http://www.foe.cymru/sites/default/files/energy_review_wales.pdf.

———. *Submission to the Welsh Affairs Committee for Its Inquiry on the Future of Nuclear Power in Wales*. March 2016. http://www.foe.cymru/sites/default/files/WAC%20Nuclear.pdf.

FORATOM. *FORATOM's Further Commentary on PINC 2016*. November 20, 2016. https://www.foratom.org/publications/#position_papers.

———. "Reaction to the 'Clean Energy for All Europeans' package." Position paper April 18, 2017. https://www.foratom.org/press-release/foratom-reacts-clean-energy-europeans-package/.

BIBLIOGRAPHY

———. *What People Really Think About Nuclear Energy.* January 2017. Brussels, Belgium: Foratom. https://www.foratom.org/publications/#topical_publications.

Forster, Peter, and Bernard Donoughue. *The Papal Encyclical: A Critical Christian Response.* GWPF Briefing 20. Global Warming Policy Foundation, 2015. http://www.thegwpf.org/content/uploads/2015/07/Forster-Donoughue1.pdf.

Foster, Claire. *Sharing God's Planet: A Christian Vision for a Sustainable Future.* Commissioned by the Church of England Mission and Public Affairs Council. London: Church House, 2005.

Fox, Tim. "UK Plutonium, CANDU with CANMOX?" IMechE, March 11, 2015. http://www.imeche.org/news/news-article/uk-plutonium-candu-with-canmox-.

Francis, Alys. "Why Are Bhopal Survivors Still Fighting for Compensation?" *BBC News*, December 2, 2014. http://www.bbc.co.uk/news/world-asia-india-30205140.

Francis, Pope. *Laudato Si'* (Encyclical Letter on Care for Our Common Home). Vatican, May 24, 2015. http://w2.vatican.va/content/francesco/en/encyclicals/documents/papa-francesco_20150524_enciclica-laudato-si.html.

———. "Show Mercy to Our Common Home." Message for the celebration of the World Day of Prayer for the Care of Creation, September 1, 2016. http://w2.vatican.va/content/francesco/en/messages/pont-messages/2016/documents/papa-francesco_20160901_messaggio-giornata-cura-creato.html.

Freeberg, Ernest. *The Age of Edison: Electric Light and the Invention of Modern America.* New York: Penguin, 2014.

FSA (Food Standards Agency). "Irradiated food." April 26, 2012. https://www.food.gov.uk/science/irradfoodqa.

———. *Radioactivity in Bottled Water.* Food Survey Information Sheet 01/14. 2014. https://www.food.gov.uk/sites/default/files/fsis-01-14-radioactivity-in-bottled-water.pdf.

———. "Radioactivity in Food: Your Questions Answered." April 18, 2011. https://www.food.gov.uk/science/rad_in_food/radioactivity.

Funk, Cary, and Lee Rainie. *Public and Scientists' Views on Science and Society.* Pew Research Center, January 29, 2015. http://www.pewinternet.org/2015/01/29/public-and-scientists-views-on-science-and-society/.

G20 2016 China. *G20 Energy Ministerial Meeting Beijing Communiqué.* Final version. https://ec.europa.eu/energy/sites/ener/files/documents/Beijing%20Communique.pdf.

Gardiner, Stephen. *The Need for a Public "Explosion" in the Principles of Nuclear Ethics.* Presented at the 2nd International Symposium on the Ethics of Environmental Health, České Budějovice, Czech Republic, June 2014. http://www.icrp.org/docs/6/27.%20Need%20for%20Public%20Explosion%20in%20the%20Principles%20of%20Nuclear%20Ethics-Gardiner.pdf.

Garner, Joel, et al. "NORM in the East Midlands' Oil and Gas Producing Region of the UK." *Journal of Environmental Radioactivity* 150 (2015) 49–56. http://www.sciencedirect.com/science/article/pii/S0265931X15300588.

Gaspar, Miklos, and Hugo Cohen Albertini. "Water Protection Measures and Community Involvement Increase Sustainability of Uranium Mining in Tanzania." *IAEA Bulletin* 56.1 (March 2015) 28–29. https://www.iaea.org/publications/magazines/bulletin/56-1.

Giddens, Anthony. *The Politics of Climate Change.* Cambridge: Polity, 2009.

BIBLIOGRAPHY

Global Nexus Initiative. *Evolving Nuclear Governance for a New Era: Policy Memo and Recommendations.* April 2017. Report based on discussions of the Global Nexus Initiative (GNI) Working Group at its June 2016 workshop in Washington, D.C. http://globalnexusinitiative.org/wp-content/uploads/2017/04/GNI-Policy-Memo-3.pdf.

———. *Nuclear Power for the Next Generation: Addressing Energy, Climate, and Security Challenges.* Full Report. April 2017 NEI and Partnership for Global Security. http://globalnexusinitiative.org/wp-content/uploads/2017/04/GNI-Digital-Report-single.pdf. (For the Executive Summary see http://globalnexusinitiative.org/wp-content/uploads/2017/04/GNI_Executive-Summary.pdf.)

Godrej, Dinyar. *The No-Nonsense Guide to Climate Change.* 3rd ed. London: New Internationalist, 2006.

Green Party of England and Wales. *For the Common Good: General Election Manifesto 2015.* 2015. https://www.greenparty.org.uk/assets/files/manifesto/Green_Party_2015_General_Election_Manifesto_Searchable.pdf.

———. "Getting Britain Working: Saving and Generating Clean Energy." Amended April 2016. https://policy.greenparty.org.uk/ey.html.

Greenpeace. *Climate Change – Nuclear Not the Answer.* April 30, 2007. http://www.greenpeace.org/international/Global/international/planet-2/report/2007/4/briefing-nuclear-not-answer-apr07.pdf.

———. "End the Nuclear Age." http://www.greenpeace.org/international/en/campaigns/nuclear/.

———. *Renewable Energy vs Nuclear Power.* February 2013. http://www.greenpeace.org/international/Global/international/publications/nuclear/2012/Fukushima/Fact%20Sheets/Renewable_Energy.pdf.

———. *2005 Energy Review – Blair Sinks Renewables and Spins Nuclear.* http://www.greenpeace.org.uk/MultimediaFiles/Live/FullReport/7343.pdf.

Grimston, Malcolm, et al. "The Siting of UK Nuclear Reactors." *Journal of Radiological Protection* 34.2 (2014) R1–R24. http://iopscience.iop.org/article/10.1088/0952-4746/34/2/R1.

Gushee, David. "Environmental Ethics: Bringing Creation Care Down to Earth." In *Keeping God's Earth: The Global Environment in Biblical Perspective*, edited by Noah J. Toly and Daniel I. Block, 245–65. Nottingham: Apollos, 2010.

Haines, Melvyn, and Nick Gaines. "Cyber Hardness of Nuclear I&C Systems." *Nuclear Future* 9.3 (May/June 2013) 53–57.

Halestrap, Andrew. *Thinking About . . . Science as a Christian Vocation: How Do We Approach a Career in Science as a Christian?* Thinking About . . . leaflet series. Christians in Science, 2016. http://www.cis.org.uk/resources/thinking/.

Hamm, Julian. "Radiation Misconceptions and Public Fears." Presented at the Nuclear Academics Meeting, Leeds, September 2, 2014. http://www.nuclearuniversities.ac.uk/pdfs/2014Presentations/Hamm_UKNADM_2014.pdf.

Han, Eun Ok, et al. "Korean Students' Behavioral Change Toward Nuclear Power Generation Through Education." *Nuclear Engineering and Technology* 46.5 (October 2014) 707–18. http://www.koreascience.or.kr/article/ArticleFullRecord.jsp?cn=OJRHBJ_2014_v46n5_707.

Hansen, James, et al. "Assessing 'Dangerous Climate Change': Required Reduction of Carbon Emissions to Protect Young People, Future Generations and Nature."

PLoS ONE 8.12 (December 2013) 1–26. http://journals.plos.org/plosone/article?id=10.1371/journal.pone.0081648.

Hansson, Sven Ove. "Should We Protect the Most Sensitive People?" *Journal of Radiological Protection* 29.2 (2009) 211–18. http://iopscience.iop.org/article/10.1088/0952-4746/29/2/008.

Harding, Sophie. *For Tomorrow Too: Living Responsibly in a World of Climate Change.* Teddington: Tearfund, 2005.

Hayhoe, Katharine. "Climate Change: Facts, Fictions, and the Christian Faith." Video of the Third Annual V. Elving Anderson Lecture in Science and Religion, University of Minnesota, April 26, 2016. https://www.youtube.com/watch?v=NqjBioAQaMo.

———. "Biography." *Katharine Hayhoe: Climate Scientiest* (blog). http://katharinehayhoe.com/wp2016/biography/.

———. "The Truth about Climate Change." Interview by John Hull. *100 Huntley Street.* Crossroads Christian Communications, May 23, 2016. http://www.100huntley.com/watch?id=223432&title=the-truth-about-climate-change.

HCBEISC (House of Commons Business, Energy and Industrial Strategy Committee). *Leaving the EU: Negotiation Priorities for Energy and Climate Change Policy.* Fourth Report of Session 2016–17. HC 909. May 2, 2017. https://www.publications.parliament.uk/pa/cm201617/cmselect/cmbeis/909/909.pdf.

———. *Oral Evidence: Leaving the EU: Negotiation Priorities for Energy and Climate Change Policy, HC 909.* February 28, 2017, published March 2, 2017. http://data.parliament.uk/writtenevidence/committeeevidence.svc/evidencedocument/business-energy-and-industrial-strategy-committee/leaving-the-eu-energy-and-climate-negotiation-priorities/oral/48479.html. (See the video http://www.parliamentlive.tv/Event/Index/54352a9b-ac49-44a8-8fab-25c533571ca6.)

HCECCC (House of Commons Energy and Climate Change Committee). *The Energy Revolution and Future Challenges for UK Energy and Climate Change Policy.* Third Report of Session 2016–17. HC 705. October 15, 2016. https://www.publications.parliament.uk/pa/cm201617/cmselect/cmenergy/705/705.pdf.

———. *Small Nuclear Power.* Fourth Report of Session 2014–15. HC 347. London, TSO, December 17, 2014. http://www.publications.parliament.uk/pa/cm201415/cmselect/cmenergy/347/347.pdf.

HCHLJC (House of Commons and House of Lords Joint Committee on the National Security Strategy). *The Next National Security Strategy.* First Report of Session 2014–15. HL 114/HC 749. London: TSO, March 3, 2015. http://www.publications.parliament.uk/pa/jt201415/jtselect/jtnatsec/114/114.pdf.

HCSTC (House of Commons Science and Technology Committee). *Devil's Bargain? Energy Risks and the Public.* First Report of Session 2012–13. Vol. 1 of 2. HC 428. London: TSO, July 9, 2012. http://www.publications.parliament.uk/pa/cm201213/cmselect/cmsctech/428/428.pdf.

HCWAC (House of Commons Welsh Affairs Committee). *The Future of Nuclear Power in Wales.* Second Report of Session 2016–17. HC 129. July 26, 2016. http://www.publications.parliament.uk/pa/cm201617/cmselect/cmwelaf/129/129.pdf.

———. "The Future of Nuclear Power in Wales." Video with witness Doug Parr, Chief Scientist of Greenpeace UK. March 21, 2016. http://parliamentlive.tv/Event/Index/979a2c34-3940-4eca-b94b-dcd6682b628b.

Hennessey, Peter. *Distilling the Frenzy: Writing the History of One's Own Times.* London: Biteback, 2012.

BIBLIOGRAPHY

Hevey, David. *Review of Public Information Programmes to Enhance Home Radon Screening Uptake and Home Remediation*. EPA Research Report 170. 2015-HW-SS-1. EPA (Ireland), March 2016. http://www.epa.ie/pubs/reports/radiation/Research_170_wrapped.pdf.

HMG (Her Majesty's Government). *The Environmental Permitting (England and Wales) Regulations 2010*. Statutory Instrument 2010 no. 675. HMSO, March 2010.

———. *The United Kingdom's Exit from and New Partnership with the European Union*. Cm 9417, February 2017. https://www.gov.uk/government/uploads/system/uploads/attachment_data/file/589191/The_United_Kingdoms_exit_from_and_partnership_with_the_EU_Web.pdf.

HLSTC (House of Lords Science and Technology Committee). *Nuclear Research and Development Capabilities*. Third Report of Sesson 2010–12. HL 221. London: TSO, November 22, 2011. http://www.publications.parliament.uk/pa/ld201012/ldselect/ldsctech/221/221.pdf.

———. *Nuclear Research and Technology: Breaking the Cycle of Indecision*. 3rd Report of Session 2016-17. HL 160. May 2, 2017. https://www.publications.parliament.uk/pa/ld201617/ldselect/ldsctech/160/160.pdf.

———. *Priorities for Nuclear Research and Technologies*. Video of evidence presented Tuesday 21 February 2017 Meeting started at 10.43am by nuclear industry witnesses. http://parliamentlive. tv/Event/index/17ed7c54-3ed9-42fb-9fcc-e56a57af3555.

Ho, Jung-Chun, et al. "Risk Perception, Trust, and Other Factors Related to a Planned New Nuclear Power Plant in Taiwan After the 2011 Fukushima Disaster." *Journal of Radiological Protection* 33.4 (2013) 773–89. http://iopscience.iop.org/article/10.1088/0952-4746/33/4/773.

Hodson, Margot R. "Editorial: Discovering a Robust Hope for Life on a Fragile Planet." *Anvil* 29.1 (2013) 1–6. https://www.degruyter.com/view/j/anv.2013.29.issue-1/anv-2013-0001/anv-2013-0001.xml.

———. *Uncovering Isaiah's Environmental Ethics*. Grove Ethics Series E161. Cambridge, UK: Grove, 2011.

Hodson, Martin J. *Donald Trump, the Environment and the Church*. JRI Briefing Paper 32. John Ray Initiative, January 20, 2017. http://www.jri.org.uk/wp/wp-content/uploads/BP32_Hodson_Trump_Environment_Church.pdf.

———. "Losing Hope? The Environmental Crisis Today." *Anvil* 29.1 (2013) 7–23. http://www.degruyter.com/view/j/anv.2013.29.issue-1/issue-files/anv.2013.29.issue-1.xml.

Hodson, Martin J., and Margot R. Hodson. *The Ethics of Climate Skepticism*. Grove Ethics Series E177. Cambridge, UK: Grove, 2015.

Hodson Martin J., and Elizabeth A. C. Rushton. *Faith, Environmental Values and Understanding: A Case Study Involving Church of England Ordinands*. JRI Briefing Paper 25. John Ray Initiative, n.d. http://www.jri.org.uk/wp/wp-content/uploads/Rushton-Hodson-JRI-briefing-25.pdf.

Holder, Rodney D. *Is the Universe Designed?* Faraday Paper 10. Faraday Institute for Science and Religion, St. Edmund's College, University of Cambridge, April 2007. http://www.faraday.st-edmunds.cam.ac.uk/resources/Faraday%20Papers/Faraday%20Paper%2010%20Holder_EN.pdf.

———. *Thinking About . . . Fine Tuning: What Does It Mean for Our Universe to Be Fine-Tuned?* Thinking About . . . leaflet series. Christians in Science, 2016. http://www.cis.org.uk/resources/thinking/.

BIBLIOGRAPHY

Hore-Lacy, Ian. Interview by Stephen Crittenden. *The Religion Report*, Radio National, September 13, 2006. http://www.abc.net.au/radionational/programs/religionreport/ian-hore-lacy/3346658.

———. *Nuclear Energy in the 21st Century*. World Nuclear University Primer. 3rd ed. London: World Nuclear University Press, 2012.

———. *Responsible Dominion: A Christian Approach to Sustainable Development*. Vancouver: Regent College Publishing, 2006.

Houghton, John. "The Changing Global Climate: Evidence, Impacts, Adaptation and Abatement." In *Keeping God's Earth: The Global Environment in Biblical Perspective*, edited by Noah J. Toly and Daniel I. Block, 187–215. Nottingham: Apollos, 2010.

———. *Global Warming, Climate Change and Sustainability: Challenge to Scientists, Policy Makers and Christians*. 3rd ed. JRI Briefing Paper 14. John Ray Initiative, January 2009. http://www.jri.org.uk/brief/Briefing_14_3rd_edition.pdf.

———. *Global Warming: The Complete Briefing*. 5th ed. Cambridge: Cambridge University Press, 2015.

———. "Q&A: Sir John Houghton." Interview by Jonathan Langley. *Mission Catalyst* 3 (2014) 4–5. http://www.bmsworldmission.org/engagecatalyst/mission-catalyst-climate-change-is-boring/qa-sir-john-houghton.

———. "Sustainable Climate and the Future of Energy." In *Creation in Crisis: Christian Perspectives on Sustainability*, edited by Robert S. White, 11–33. London: SPCK, 2009.

Houghon, John, with Gill Tavner. *In the Eye of the Storm: The Autobiography of Sir John Houghton*. Oxford: Lion, 2013.

HPA (Health Protection Agency). *Radon and Public Health: Report of the Independent Advisory Group on Ionising Radiation*. RCE-11. June 2009. http://webarchive.nationalarchives.gov.uk/20110109132023/http://hpa.org.uk/web/hpawebfile/hpaweb_c/1243838496865.

Hulme, Philippa Gardom, and Hillary Taunton. *GCSE Physics: Revision Guide*. Twenty First Century Science. Oxford: Oxford University Press and OCR, 2012.

Hutchings, David, and Tom McLeish. *Let There Be Science: Why God Loves Science, and Science Needs God*. Oxford: Lion, 2017.

Hutchinson, Ian, and Leith Anderson. "Being a Christian and a Scientist." *Today's Conversation* (podcast), NAE, October 15, 2015. http://nae.net/hutchinsonpodcast/.

IAEA (International Atomic Energy Agency). *Atoms for Peace and Development: How the IAEA Supports the Sustainable Development Goals*. 2015. https://www.iaea.org/sites/default/files/sdg-brochure_forweb.pdf.

———. *Climate Change and Nuclear Power 2015*. September 2015. http://www-pub.iaea.org/MTCD/Publications/PDF/CCANP2015Web-78834554.pdf.

———. *Extent of Environmental Contamination by Naturally Occurring Radioactive Material (NORM) and Technological Options for Mitigation*. Technical Report Series 419. 2003. http://www-pub.iaea.org/MTCD/Publications/PDF/TRS419_web.pdf.

———. *IAEA Annual Report 2014*. GC(59)/7. 2015. https://www.iaea.org/sites/default/files/gc59-7_en.pdf.

———. *IAEA Safeguards: Serving Nuclear Non-Proliferation*. June 2015. https://www.iaea.org/sites/default/files/safeguards_web_june_2015_1.pdf.

———. *IAEA Safety Glossary: Terminology Used in Nuclear Safety and Radiation Protection*. 2007 ed. http://www-pub.iaea.org/MTCD/publications/PDF/Pub1290_web.pdf.

———. "IAEA Values." January 26, 2016. https://www.iaea.org/about/employment/values.

———. *INES: The International Nuclear and Radiological Event Scale.* 08-26941/E. n.d. https://www.iaea.org/sites/default/files/ines.pdf.

———. "International Conference on Nuclear Security: Commitments and Actions." Vienna, December 5–9, 2016. Conference ID: 50809 (CN-244). http://www-pub.iaea.org/iaeameetings/50809/International-Conference-on-Nuclear-Security-Commitments-and-Actions.

———. *Linking Nuclear Power and Environment: Safe, Secure, Sustainable Nuclear Power.* 2013. https://www.iaea.org/sites/default/files/np0613.pdf.

———. *The Radiological Accident in Goiânia.* STI/PUB/815. 1988. http://www-pub.iaea.org/MTCD/publications/PDF/Pub815_web.pdf.

———. *Report of the Operational Safety Review Team (OSART) Mission to Sizewell B Nuclear Power Station 5–22 October 2016.* IAEA–NSNI/OSART/015/185. https://www.gov.uk/government/uploads/system/uploads/attachment_data/file/540260/Sizewell_B_final_report2.pdf.

———. *Status of Small Reactor Designs Without On-Site Refuelling.* IAEA TECDOC 1536. 2007. http://www-pub.iaea.org/books/IAEABooks/7658/Status-of-Small-Reactor-Designs-Without-On-Site-Refuelling.

IEA (International Energy Agency). *Energy and Air Pollution: Executive Summary. World Energy Outlook Special Report.* OECD and IEA, 2016. http://www.iea.org/publications/freepublications/publication/WorldEnergyOutlookSpecialReportEnergyandAirPollution_Executivesummary_EnglishVersion.pdf.

———. *Energy and Climate Change: World Energy Outlook Special Report.* OECD and IEA, 2015. https://www.iea.org/publications/freepublications/publication/WEO2015SpecialReportonEnergyandClimateChange.pdf.

———. *Energy Policies of IEA Countries: Belgium. 2016 Review.* OECD and IEA, 2016. http://www.iea.org/publications/freepublications/publication/Energy_Policies_of_IEA_Countries_Belgium_2016_Review.pdf.

———. *Ensuring Green Growth in a Time of Economic Crisis: The Role of Energy Technology.* 2009. http://www.iea.org/publications/freepublications/publication/ensuring_green_growth.pdf.

———. "IEA Executive Director Delivers Keynote Address at World Nuclear Exhibition." June 28, 2016. http://www.iea.org/newsroom/news/2016/june/iea-executive-director-delivers-keynote-address-at-world-nuclear-exhibition.html.

———. "Presentation: Launch of the IEA World Energy Outlook Special Report on Energy and Air Pollution." Video of presentation to the press, London, June 27, 2016. https://www.youtube.com/watch?v=taXDge83LoA.

———. *Tracking Clean Energy Progress 2016: Energy Technology Perspectives 2016 Excerpt. IEA Input to the Clean Energy Ministerial.* OECD and IEA, 2016. http://www.iea.org/publications/freepublications/publication/TrackingCleanEnergyProgress2016.pdf.

IMechE (Institution of Mechanical Engineers). *Engineering the UK Electricity Gap.* January 2016. http://www.imeche.org/docs/default-source/position-statements-energy/imeche-ps-electricity-gap.pdf.

———. *Leaving the EU: The Euratom Treaty.* February 2017. http://www.imeche.org/news/news-article/uk-s-break-from-eu-could-threaten-nuclear-fuel-supplies-and-reactor-build-report-says.

BIBLIOGRAPHY

———. "New Survey: Majority of the Public Support UK Nuclear Power." October 13, 2015. https://www.imeche.org/news/news-article/new-survey-majority-of-the-public-support-uk-nuclear-power.

———. *Nuclear Build: A Vote of No Confidence?* March 2010. http://www.imeche.org/docs/default-source/1-oscar/reports-policy-statements-and-documents/nuclear-build---a-vote-of-no-confidence.pdf.

INPO (Institute of Nuclear Power Operations). *Traits of a Healthy Nuclear Safety Culture.* INPO 12–012. Rev. 1. April 2013. http://nuclearsafety.info/wp-content/uploads/2010/07/Traits-of-a-Healthy-Nuclear-Safety-Culture-INPO-12-012-rev.1-Apr2013.pdf.

Interfaith Power & Light. "Nuclear." June 24, 2016. http://www.interfaithpowerandlight.org/question/nuclear.

———. "17 Anglican Bishops from All Six Continents Have Called for Urgent Prayer and Action on the 'Unprecedented Climate Crisis.'" March 30, 2015. http://www.interfaithpowerandlight.org/2015/03/17-anglican-bishops-from-all-six-continents-have-called-for-urgent-prayer-and-action-on-the-unprecedented-climate-crisis/.

IPCC (Intergovernmental Panel on Climate Change). *Climate Change 2014: Mitigation of Climate Change. Contribution of Working Group III to the Fifth Assessment Report of the Intergovernmental Panel on Climate Change.* Edited by Ottmar Edenhofer et al. Cambridge: Cambridge University Press, 2014. http://www.ipcc.ch/report/ar5/wg3/. For a helpful summary, see the section "Summary for Policymakers" at http://www.ipcc.ch/pdf/assessment-report/ar5/wg3/ipcc_wg3_ar5_summary-for-policymakers.pdf.

Ipsos MORI. *Public Attitudes to Science 2014.* Report for the Department for Business, Innovation and Skills. URN BIS/14/P111. March 2014. https://www.gov.uk/government/uploads/system/uploads/attachment_data/file/348830/bis-14-p111-public-attitudes-to-science-2014-main.pdf.

Irvine, Maxwell. *Nuclear Power: A Very Short Introduction.* Oxford: Oxford University Press, 2011.

Ivanov, V. K., et al. "Estimating the Lifetime Risk of Cancer Associated with Multiple CT Scans." *Journal of Radiological Protection* 34.4 (2015) 825–41. http://iopscience.iop.org/article/10.1088/0952-4746/34/4/825.

John Ray Initiative. "A Sustainable Future?" Conference March 5, 2016. http://www.jri.org.uk/%20events/a-sustainable-future-jri-conference-5-march-2016/.

Jones, Steve. "Health Effects of the Windscale Pile Fire." Invited editorial. *Journal of Radiological Protection* 36.4 (2016) E23–E25. http://iopscience.iop.org/article/10.1088/0952-4746/36/4/E23/pdf.

Juvenal. *The Sixteen Satires.* Translated by Peter Green. Rev. ed. London: Penguin, 1974.

Keefe, Molly, et al. *Safety Culture Common Language.* NUREG-2165. US Nuclear Regulatory Commission, March 2014. http://www.nrc.gov/docs/ML1408/ML14083A200.pdf.

Kemeny Commission. *Report of the President's Commission on the Accident at Three Mile Island. The Need for Change: The Legacy of TMI.* October 1979. Washington, DC: US Government Printing Office, 1979. http://www.threemileisland.org/downloads/188.pdf.

Kennedy, Robert F., Jr. *Crimes Against Nature: Standing Up to Bush and the Kyoto Killers Who Are Cashing In on Our World.* London: Penguin, 2005.

Kerrigan, David. "Bored to Death." *Mission Catalyst* 3 (2014) 2. http://www.bmsworldmission.org/engagecatalyst/mission-catalyst-climate-change-is-boring/bored-death-climate-change-editorial.

Kharecha, Pushker, and James E. Hansen. "Coal and Gas Are Far More Harmful than Nuclear Power." NASA Goddard Institute for Space Studies, Science Briefs, April 2013. http://www.giss.nasa.gov/research/briefs/kharecha_02/.

———. "Prevented Mortality and Greenhouse Gas Emissions from Historical and Projected Nuclear Power." *Environmental Science & Technology* 47.9 (2013) 4889–95. http://pubs.acs.org/doi/abs/10.1021/es3051197.

———. "Response to Comment by Rabilloud on 'Prevented Mortality and Greenhouse Gas Emissions from Historical and Projected Nuclear Power.'" *Environmental Science and Technology* 47 (2013) 13900–901. http://pubs.giss.nasa.gov/docs/2013/2013_Kharecha_Hansen_3.pdf.

———. "Response to Comment on 'Prevented Mortality and Greenhouse Gas Emissions from Historical and Projected Nuclear Power.'" *Environmental Science amd Technology* 47 (2013) 6718–19. https://pubs.giss.nasa.gov/docs/2013/2013_Kharecha_kho6000s.pdf.

Kidd, Steve. "Is Climate Change the Worst Argument for Nuclear?" *Nuclear Engineering International*, January 21, 2015. http://www.neimagazine.com/opinion/opinionis-climate-change-the-worst-argument-for-nuclear-4493537/.

———. "The Nuclear Establishment – What Lies Within?" *Nuclear Engineering International*, February 8, 2017. http://www.neimagazine.com/opinion/opinionthe-nuclear-establishment-what-lies-within-5735341/.

King, David. "Renewable Energy and Nuclear Power for the UK." Video of the Lord Lewis Prize Lecture to the Royal Society of Chemistry, London, March 5, 2013. https://www.youtube.com/watch?v=nPQFQDIg3QM.

Kirner, Nancy. "The WARP Report: Wow!" *Health Physics News* 44.2 (February 2016) 4. http://hps.org/hpspublications/newsletter_vol44n002.pdf.

Kohzaki, Masaoki, et al. "What Have We Learned from a Questionnaire Survey of Citizens and Doctors Both Inside and Outside Fukushima?: Survey Comparison Between 2011 and 2013." *Journal of Radiological Protection* 35.1 (2015) N1–N17. http://iopscience.iop.org/article/10.1088/0952-4746/35/1/N1.

König, C., et al. "Remediation of TENORM Residues: Risk Communication in Practice." *Journal of Radiological Protection* 34.3 (2014) 575–93. http://iopscience.iop.org/article/10.1088/0952-4746/34/3/575.

Korsnick, Maria. *Nuclear Power Is Critical Infrastructure, Maria Korsnick, Feb. 9, 2017.* 2017 Wall Street Briefing by Maria Korsnick President and Chief Executive Officer NEI February 9, 2017. https://www.nei.org/News-Media/Speeches/Nuclear-Power-is-Critical-Infrastructure. (Briefing given for the NEI "Annual Update on Nuclear Energy in America" and is at: https://www.youtube.com/watch?v=3LOzmdENAns.)

Labour Party. *Britain Can Be Better: The Labour Party Manifesto 2015.* http://www.labour.org.uk/page/-/BritainCanBeBetter-TheLabourPartyManifesto2015.pdf.

Lausanne Movement. *The Cape Town Commitment: A Confession of Faith and a Call to Action.* 2011. https://www.lausanne.org/content/ctc/ctcommitment.

Leech, Jonathan, and Rupert Cowen. "Brexit White Paper Confuses Euratom Debate." *World Nuclear News*, February 8, 2017. http://www.world-nuclear-news.org/V-Brexit-white-paper-confuses-Euratom-debate-08021702.html.

BIBLIOGRAPHY

Legates, David R., and G. Cornelis van Kooten. *A Call to Truth, Prudence, and Protection of the Poor 2014: The Case Against Harmful Climate Policies Gets Stronger*. Cornwall Alliance for the Stewardship of Creation, September 2014. http://www.cornwallalliance.org/wp-content/uploads/2014/09/A-Call-to-Truth-Prudence-and-Protection-of-the-Poor-2014-The-Case-Against-Harmful-Climate-Policies-Gets-Stronger.pdf.

Liberal Democrats. *Stronger Economy, Fairer Society: Opportunity for Everyone. Liberal Democrat Manifesto 2015*. http://www.libdems.org.uk/manifesto-clear-print.

Lindo, Gerald. "Differences in Climate Impacts Between 1.5°C and 2°C." IPCC presentation at the UNFCCC, Bonn, Germany, May 16–26, 2016. http://unfccc.int/files/science/workstreams/research/application/pdf/part1.5_aosis_lindo.pdf.

Lothes Biviano, Erin, et al. "Catholic Moral Traditions and Energy Ethics for the Twenty-First Century." *Journal of Moral Theology* 5.2 (2016) 1–36. http://msmary.edu/College_of_liberal_arts/department-of-theology/jmt-files/Energy%20Ethics.pdf.

Lovelock, James. *The Revenge of Gaia: Why the Earth Is Fighting Back—and How We Can Still Save Humanity*. London: Allen Lane, 2006.

Lovering, Jessica R., et al. "Historical Construction Costs of Global Nuclear Power Reactors." *Energy Policy* 91 (April 2016) 371–82. http://www.sciencedirect.com/science/article/pii/S0301421516300106.

Lucas, Caroline. *Honourable Friends?: Parliament and the Fight for Change*. London: Portobello, 2015.

Lucas, Edward. *Cyberphobia: Identity, Trust, Security and the Internet*. London: Bloomsbury, 2015. http://www.edwardlucas.com/.

Lucas, Ernest. *Can We Believe Genesis Today?: The Bible and the Questions of Science*. 3rd ed. Nottingham: Inter-Varsity, 2005.

———. *Interpreting Genesis in the 21st Century*. Faraday Paper 11. Faraday Institute for Science and Religion, St. Edmund's College, University of Cambridge, 2007. http://www.faraday.st-edmunds.cam.ac.uk/resources/Faraday%20Papers/Faraday%20Paper%2011%20Lucas_EN.pdf.

Lynas, Mark. "Nuclear Power: A Convert." *New Statesman*, May 30, 2005. http://www.newstatesman.com/node/195308.

———. "Nuclear Power Support from Former Sceptic Mark Lynas." Video. *BBC News*, October 8, 2013. http://www.bbc.co.uk/news/uk-politics-24445371.

———. *Nuclear 2.0: Why a Green Future Needs Nuclear Power*. Cambridge: UIT, 2013.

———. *Six Degrees: Our Future on a Hotter Planet*. London: Harper Perennial, 2008.

Ma'anit, Adam. "Nuclear Is the New Black." *New Internationalist*, September 2005. http://newint.org/features/2005/09/01/keynote/.

MacKay, David J. C. "David MacKay—Last Interview and Tribute." Video of interview by Mark Lynas, April 3, 2016. Posted April 27, 2016. http://www.marklynas.org/2016/04/david-mackay-last-interview-tribute/.

———. *Sustainable Energy—Without the Hot Air*. Cambridge: UIT, 2009. http://www.withouthotair.com.

MacKerron, Gordon. *Evaluation of Nuclear Decommissioning and Waste Management*. URN 12D/002/. University of Sussex, March 2012. http://sro.sussex.ac.uk/40037/1/Evaluation_of_nuclear_decommissioning_and_waste_management.pdf.

Markandya, Anil, and Paul Wilkinson. "Electricity Generation and Health." *Lancet* 370 (2007) 979–90. http://www.bigthunderwindpower.ca/files/resources/Electricity_generation_and_health_(The_Lancet_2007).pdf.

Marr, Andrew. *A History of Modern Britain*. London: Pan, 2008.
Marshall, George. *Don't Even Think About It: Why Our Brains Are Wired to Ignore Climate Change*. London: Bloomsbury, 2015.
Martin, Alan D., et al. *An Introduction to Radiation Protection*. 6th ed. London: Hodder Arnold, 2012.
May, Ernest R. "John F Kennedy and the Cuban Missile Crisis." BBC, November 18, 2013. http://www.bbc.co.uk/history/worldwars/coldwar/kennedy_cuban_missile_01.shtml.
May, Michael M. "Safety First: The Future of Nuclear Energy Outside the United States." *Bulletin of the Atomic Scientists* 73.1 (2017) 38–43. Published online December 13, 2016. Special Issue: Should Nuclear Power Be a Major Part of the World's Response to Climate Change. http://www.tandfonline.com/doi/full/10.1080/00963402.2016.1264210.
McGeoghegan, D., et al. "Mortality and Cancer Registration Experience of the Sellafield Workers Known to Have Been Involved in the 1957 Windscale Accident: 50 Year Follow-Up." *Journal of Radiological Protection* 30.3 (2010) 407–31. http://iopscience.iop.org/article/10.1088/0952-4746/30/3/001.
McGowan, D. R., et al. "Iodine-131 Monitoring in Sewage Plant Outflow." *Journal of Radiological Protection* 34.1 (2014) 1–14. http://iopscience.iop.org/article/10.1088/0952-4746/34/1/1.
McGuire, Bill. *Waking the Giant: How a Changing Climate Triggers Earthquakes, Tsunamis, and Volcanoes*. Oxford: Oxford University Press, 2012.
McKenna, Josephine. "Pope Francis Says Destroying the Environment Is a Sin." *Guardian*, September 1, 2016. https://www.theguardian.com/world/2016/sep/01/pope-francis-calls-on-christians-to-embrace-green-agenda.
McKeown, James. *Genesis*. Two Horizons Old Testament Commentary. Grand Rapids: Eerdmans, 2008.
McLeish, Tom. "Faith and Wisdom in Science." Video interview with Eleanor Puttock at the Faraday Institute Summer Course No 9: Science and Religion—Engaging in Constructive Dialogue, October 30, 2014. https://www.youtube.com/ watch?v=C8X6p17ttQi.
McNally, Richard J. Q., et al. "A Geographical Study of Thyroid Cancer Incidence in North-West England Following the Windscale Nuclear Reactor Fire of 1957." *Journal of Radiological Protection* 36.4 (December 2016) 934–52. http://iopscience.iop.org/article/10.1088/0952-4746/36/4/934/meta.
Mecklin, John. "Introduction: Nuclear Power and the Urgent Threat of Climate Change." *Bulletin of the Atomic Scientists* 73.1 (2017) 1. Published online December 22, 2016. Special Issue: Should Nuclear Power Be a Major Part of the World's Response to Climate Change. http://www.tandfonline.com/doi/full/10.1080/00963402.2016.1265351.
Mellen, Andy, and Neil Hollow. *No Oil in the Lamp: Fuel, Faith and the Energy Crisis*. London: Darton, Longman and Todd, 2012.
Ming, Zeng, et al. "Nuclear Energy in the Post-Fukushima Era: Research on the Developments of the Chinese and Worldwide Nuclear Power Industries." *Renewable and Sustainable Energy Reviews* 58 (May 2016) 147–56. http://www.sciencedirect.com/science/article/pii/S1364032115015488.
Mobbs, S., et al. *An Introduction to the Estimation of Risks Arising from Exposure to Low Doses of Ionising Radiation*. HPA-RPD-055. HPA, June 2009. https://www.

gov.uk/government/uploads/system/uploads/attachment_data/file/340094/HPA-RPD-055_for_website.pdf.

Monbiot, George. "Going Critical: How the Fukushima Disaster Taught Me to Stop Worrying and Embrace Nuclear Power." March 21, 2011. http://www.monbiot.com/2011/03/21/going-critical/. Also published in *Guardian*, March 22, 2011.

———. "Power Crazed: Why Do We Transfer the Real Health Risks Inflicted by Coal onto Nuclear Energy?" December 16, 2013. http://www.monbiot.com/2013/12/16/power-crazed/. Also published in *Guardian*, December 16, 2013.

Montefiore, Hugh. "Hugh Montefiore: We Need Nuclear Power to Save the Planet from Looming Catastrophe." *Independent*, October 21, 2004. http://www.independent.co.uk/voices/commentators/hugh-montefiore-we-need-nuclear-power-to-save-the-planet-from-looming-catastrophe-5351398.html.

Moo, Douglas J. "Eschatology and Environmental Ethics." In *Keeping God's Earth: The Global Environment in Biblical Perspective*, edited by Noah J. Toly and Daniel I. Block, 23–43. Nottingham: Apollos, 2010.

Moore, Patrick. "Going Nuclear." *Washington Post*, April 16, 2006. http://www.washingtonpost.com/wp-dyn/content/article/2006/04/14/AR2006041401209.html.

———. "Patrick Moore on How to Stop Worrying and Love Mother Earth—Part 2." Interview by Joseph F. Cotto. August 10, 2012. http://ecosense.me/2017/01/18/issues-4/.

———. "Should We Celebrate Carbon Dioxide?" Video of the 2015 Annual GWPF Lecture, London, October 14, 2015. http://nsb.com/speakers/patrick-moore/.

Murphy, Brian, et al. *Power People: The Civil Nuclear Workforce 2009–2025*. Renaissance Nuclear Skills Series 1. Cogent SSC, 2011. http://namrc.group.shef.ac.uk/wp-content/uploads/2011/01/Cogent-PowerPeople.pdf.

NAE (National Association of Evangelicals). "About NAE." http://nae.net/about-nae/.

National Cyber Security Centre. "Common Cyber Attacks: Reducing the Impact." January 2016. https://www.ncsc.gov.uk/white-papers/common-cyber-attacks-reducing-impact.

National Grid. *Future Energy Scenarios: GB Gas and Electricity Transmission*. July 2016. http://fes.nationalgrid.com/fes-document/.

———. *Future Energy Scenarios in Five Minutes*. July 2016. http://investors.nationalgrid.com/~/media/Files/N/National-Grid-IR/reports/FES%20in%205%20minutes.pdf.

NCRP (National Council on Radiation Protection and Measurements). *Ionizing Radiation Exposure of the Population of the United States*. Report 160. 2009. https://www.ncrppublications.org/Reports/160.

———. *Where Are the Radiation Professionals (WARP)?* NCRP Statement 12. December 17, 2015. http://ncrponline.org/wp-content/themes/ncrp/PDFs/Statement_12.pdf.

NDA (Nuclear Decommissioning Authority). "About Us." https://www.gov.uk/government/organisations/nuclear-decommissioning-authority/about.

———. *Nuclear Decommissioning Authority Business Plan: 1 April 2017 to 31 March 2020*. SG/2017/24 March 2017. https://www.gov.uk/government/uploads/system/uploads/attachment_data/file/604324/NDA_Business_Plan_2017_to_2020.pdf.

———. *Nuclear Provision: Explaining the Costs of Cleaning Up Britain's Nuclear Legacy*. September 1, 2016. https://www.gov.uk/government/publications/nuclear-provision-explaining-the-cost-of-cleaning-up-britains-nuclear-legacy.

NEA (Nuclear Energy Agency). *2016 NEA Annual Report*. NEA No. 7349. OECD, 2016. https://www.oecd-nea.org/pub/activities/ar2016/ar2016.pdf.

BIBLIOGRAPHY

———. *Managing Environmental and Health Impacts of Uranium Mining*. NEA No. 7062. OECD, 2014. http://www.oecd-nea.org/ndd/pubs/2014/7062-mehium.pdf.

———. *Nuclear Energy: Combating Climate Change*. NEA No. 7208. OECD, 2015. https://www.oecd-nea.org/ndd/pubs/2015/7208-climate-change-2015.pdf.

———. *Public Attitudes to Nuclear Power*. NEA No. 6859. OECD, 2010. https://www.oecd-nea.org/ndd/reports/2010/nea6859-public-attitudes.pdf.

NEA (Nuclear Energy Agency) and IAEA (International Atomic Energy Agency). *Uranium 2016: Resources, Production and Demand*. NEA No. 7301. OECD, 2016. http://www.oecd-nea.org/ndd/pubs/2016/7301-uranium-2016.pdf.

NEI (Nuclear Energy Institute). "Japan Nuclear Update: NRA Considering Restart Applications of 26 Nuclear Plants." March 16, 2017. https://www.nei.org/News-Media/News/Japan-Nuclear-Update.

———. *NEI US Public Opinion Snapshot: Nuclear Plant Neighbors and the Public, Fall 2015*. http://www.nei.org/CorporateSite/media/filefolder/Backgrounders/Reports-Studies/Plant-Neighbors-and-the-Public-Snapshot-Fall-2015.pdf.

———. "New Mexico Interim Storage Facility Plan Complements Other Used Nuclear Fuel Options: Holtec International Submits License Application to NRC." April 5, 2017. https://www.nei.org/News-Media/Media-Room/News-Releases/New-Mexico-Interim-Storage-Facility-Plan-Complemen.

———. "New York's Support for Nuclear Preserves Clean Air, Quality Jobs." March 7, 2017. https://www.nei.org/News-Media/News/News-Archives/New-York-s-Support-for-Nuclear-Preserves-Clean-Air

———. *Nuclear Energy: Powering America's Future*. October 2016. https://www.nei.org/CorporateSite/media/filefolder/Publications-Brochures/Brochures/2016_New_Plant_Brochure.pdf.

———. "Public Opinion: Americans Voice Strong Support for Nuclear Energy." Poll conducted September to October 2016. https://www.nei.org/Knowledge-Center/Public-Opinion.

———. "Map of US Nuclear Plants." https://www.nei.org/Knowledge-Center/Map-of-US-Nuclear-Plants.

Nénot, Jean-Claude. "Radiation Accidents Over the Last 60 Years." *Journal of Radiological Protection* 29.3 (2009) 301–20. http://iopscience.iop.org/article/10.1088/0952-4746/29/3/R01.

NFCRC (Nuclear Fuel Cycle Royal Commission). *Nuclear Fuel Cycle Royal Commission Report*. May 2016. http://yoursay.sa.gov.au/system/NFCRC_Final_Report_Web.pdf.

NI (Nuclear Institute). *Brexit and the Euratom Treaty Issue*. April 2017. http://www.nuclearinst.com/write/MediaUploads/PDFs/NI_-_Response_to_Brexit_and_Euratom_-_April_2017.pdf.

———. *Nuclear Institute Response to the House of Lords Inquiry into Priorities for Nuclear Research and Technologies*. February 22, 2017. http://www.nuclearinst.com/write/MediaUploads/PDFs/NI_-_Response_to_House_of_Lords_Inquiry.pdf.

———. "Nuclear Organisations Join Forces to Bridge the Knowledge Gap." May 5, 2015. http://www.nuclearinst.com/News/nuclear-organisations-join-forces-to-bridge-the-knowledge-gap.

———. "Trailblazing Apprenticeships for Nuclear." *Nuclear Future* 11.5 (September/October 2015) 8.

NIA (Nuclear Industry Association). *Decommissioning*. Briefing paper. 2016. https://www.niauk.org/wp-content/uploads/2016/09/Decommissioning.pdf.

BIBLIOGRAPHY

———. *Exiting Euratom*. May 2017. https://www.niauk.org/wp-content/uploads/2017/05/Exiting-Euratom_May17.pdf.
———. "Exploring Female Attitudes Towards Nuclear Power." Video. July 7, 2014. https://www.niauk.org/media-centre/videos/exploring-female-attitudes-towards-nuclear-power/.
———. "Facts and Information for Nuclear Energy." 2016. http://www.niauk.org/facts-and-information-for-nuclear-energy.
———. "NIA Welcomes SMR Deployment Report." September 29, 2016. https://www.niauk.org/media-centre/press-releases/nia-welcomes-smr-deployment-report/.
———. "Nuclear Industry Commits to Public Engagement." December 3, 2015. https://www.niauk.org/media-centre/press-releases/nuclear-industry-commits-to-public-engagement/.
———. *Plutonium Management*. Briefing paper. November 2015. https://www.niauk.org/wp-content/uploads/2016/09/Plutonium-Management.pdf.
———. "Public Opinion of the Nuclear Industry." Video. July 7, 2016. https://www.niauk.org/media-centre/videos/public-opinion-of-the-nuclear-industry/.
———. *UK Advanced Boiling Water Reactor (UK ABWR) Justification Debate*. Briefing paper. April 2015. http://www.niauk.org/justification-application-uk-abwr.
———. *UK Public Opinion*. January 2016. http://www.niauk.org/images/graphics/facts_enlargements/public_op2015.pdf.
———. "Unions Back New Nuclear." July 1, 2016. https://www.niauk.org/media-centre/member-news/unions-back-new-nuclear/.
———. "Waste Management." 2016. https://www.niauk.org/industry-issues/waste-management/.
NIC (Nuclear Industry Council). *In the Public Eye: Nuclear Energy and Society*. July 2014. https://www.gov.uk/government/uploads/system/uploads/attachment_data/file/360669/In_the_Public_Eye_-_Nuclear_Energy_and_Society_-_NICJuly2014.pdf.
———. *Nuclear Energy and Society: A Concordat for Public Engagement*. December 2015. http://www.niauk.org/images/pdfs/publications/Public%20Engagement%20Concordat%20Dec15.pdf.
NIRAB (Nuclear Innovation and Research Advisory Board). *NIRAB Final Report: 2014-16*. NIRAB-117-3 February 2017. http://nirab.org.uk/media/10139/nirab-117-3-nirab-final-report_web.pdf.
NNL (National Nuclear Laboratory). *Small Modular Reactors (SMR) Feasibility Study*. December 2014. http://www.nnl.co.uk/media/1627/smr-feasibility-study-december-2014.pdf.
NNSA (National Nuclear Security Administration). *Prevent, Counter, and Respond—A Strategic Plan to Reduce Global Nuclear Threats (FY 2017–FY 2021)*. Report to Congress, March 2016. https://nnsa.energy.gov/aboutus/ourprograms/dnn/npcr.
Northcott, Michael. "Sustaining Ethical Life in the Anthropocene." In *Creation in Crisis: Christian Perspectives on Sustainability*, edited by Robert S. White, 225–40. London: SPCK, 2009.
NRC (Nuclear Regulatory Commission). *Protecting Our Nation*. NUREG/BR-0314, rev. 4. August 2015. http://www.nrc.gov/docs/ML1523/ML15232A263.pdf.
———. "Radiation and National Security." April 28, 2016. http://www.nrc.gov/about-nrc/radiation/rad-nat-security.html.

———. "Safety-Conscious Work Environment." April 1, 2016. http://www.nrc.gov/about-nrc/safety-culture/scwe.html.

———. "Safety Culture." March 3, 2016. http://www.nrc.gov/about-nrc/safety-culture.html.

———. "Safety Culture and Nuclear Reactors." April 1, 2016. http://www.nrc.gov/about-nrc/safety-culture/sc-nuclear-reactors.html.

———. "Safety Culture Policy Statement." April 1, 2016. http://www.nrc.gov/about-nrc/safety-culture/sc-policy-statement.html.

NSSG (Nuclear Skills Strategy Group). *Nuclear Skills Strategic Plan: Executive Summary*. December 2016. http://www.cogentskills.com/media/76261/skills-strategy-executive-summary.pdf.

———. *Nuclear Skills Strategic Plan: Government and Industry Working Together to Build Excellence in Nuclear Skills*. December 2016. http://www.cogentskills.com/media/76258/national-nuclear-skills-strategic-plan.pdf.

Nuclear Forum. "Discover the World's Nuclear Power Plants and Repositories with Nuclearplanet." September 8, 2016. http://www.nuklearforum.ch/de/en/nuclearplanet.

Nuclear Threat Initiative. "Outpacing Cyber Threats: Priorities for Cybersecurity at Nuclear Facilities – Executive Summary." December 7, 2016. http://www.nti.org/analysis/reports/outpacing-cyber-threats-priorities-cybersecurity-nuclear-facilities/.

NucNet. "Belgium Closures Will 'Seriously Challenge' Energy Security, Warns IEA." May 19, 2016. http://www.nucnet.org/all-the-news/2016/05/19/belgium-closures-will-seriously-challenge-energy-security-warns-iea.

———. "Global Uranium Supply 'More than Adequate' for Foreseeable Future, Says Red Book." December 1, 2016. http://www.nucnet.org/all-the-news/2016/12/01/global-uranium-supply-more-than-adequate-for-foreseeable-future-says-red-book.

———. "'Major Shift' Has Begun Towards Low-Carbon Energy Such as Nuclear, Says IEA." September 14, 2016. http://www.nucnet.org/all-the-news/2016/09/14/major-shift-has-begun-towards-low-carbon-energy-such-as-nuclear-says-iea.

———. "Referendum Result Shows Public Confidence in Nuclear, Says Swiss Industry Group." November 28, 2016. http://www.nucnet.org/all-the-news/2016/11/28/referendum-result-shows-public-confidence-in-nuclear-says-swiss-industry-group.

———. "Switzerland Rejects Plans to Speed Up Country's Nuclear Phaseout." November 27, 2016. http://www.nucnet.org/all-the-news/2016/11/27/switzerland-rejects-plans-to-speed-up-country-s-nuclear-phaseout.

———. "Switzerland Votes To Phase Out Nuclear Energy." May 22, 2017. http://www.nucnet.org/all-the-news/2017/05/22/switzerland-votes-to-phase-out-nuclear-energy.

NuScale Power. "NuScale Announces MOX Capability." January 20, 2016. http://newsroom.nuscalepower.com/press-release/company/nuscale-announces-mox-capability.

Nuttall, William J. "Britain, Nuclear Energy and the Future." Video of the IET Clerk Maxwell Lecture 2015, Royal Institution, London, March 15, 2015. https://www.youtube.com/watch?v=wCviKi5aD-I.

Nuttall, William J., and John E. Earp. "Nuclear Energy in the UK: Safety Culture and Industrial Organisation." In *New Nuclear Power Industry Procurement Markets: International Experiences*, edited by Ilchong Nam and Geoffrey Rothwell, 142–83.

BIBLIOGRAPHY

Research Monograph 2014–01. Sejong-Si, South Korea: Korean Development Institute, 2014. http://oro.open.ac.uk/44243/1/UK_Paper_v21.pdf.

Oatway, W. B., et al. *Ionising Radiation Exposure of the UK Population: 2010 Review.* PHE-CRCE-026. PHE, April 2016. https://www.gov.uk/government/publications/ionising-radiation-exposure-of-the-uk-population-2010-review.

O'Connor, C., et al. *Radiation Doses Received by the Irish Population 2014.* RPII 14/02. June 2014. http://www.epa.ie/pubs/reports/radiation/RPII_Radiation_Doses_Irish_Population_2014.pdf.

OECD and NEA. *Impacts of the Fukushima Daiichi Accident on Nuclear Development Policies.* NEA No. 7212. Paris, France: OECD/NEA, 2017. http://www.oecd-nea.org/ndd/pubs/2017/7212-impacts-fukushima-policies.pdf.

O'Flaherty, Tom. "If Use of Nuclear Power Continues, Will There Be Enough Uranium?" *Nuclear Future* 7.6 (November/December 2011) 33–41.

Olajide, Tolulope, et al., "Challenging Perceptions of STEM." *Nuclear Future* 11.2 (March/April 2015) 46–49.

ONR (Office for Nuclear Regulation). "Advanced Passive 1000—AP1000®." March 30, 2017. http://www.onr.org.uk/new-reactors/ap1000/index.htm.

———. "Annual Civil Plutonium and Uranium Figures as of 31 December 2014." November 14, 2016. http://www.onr.org.uk/safeguards/civilplut14.htm.

———. "Basic Safeguards Glossary." March 3, 2016. http://www.onr.org.uk/safeguards/glossary.htm.

———. "Civil Nuclear Security." September 21, 2016. http://www.onr.org.uk/ocns/.

———. *Enforcement Policy Statement.* ONR-ENF-POL-001 rev. 0. April 1, 2014. http://www.onr.org.uk/documents/2014/enforcement-policy-statement.pdf.

———. *Events Reported to Nuclear Safety Regulator in the Period of 1 April 2001 to 31 March 2015.* February 2016. http://news.onr.org.uk/2016/02/events-reported-to-nuclear-safety-regulator-2001-15/.

———. *Finding a Balance: Guidance on the Sensitivity of Nuclear and Related Material and Its Disclosure.* Version 3. April 2, 2014. http://www.onr.org.uk/ocns/balance.pdf.

———. "Frequently Asked Questions." September 20, 2016. http://www.onr.org.uk/new-reactors/faq.htm.

———. "Fukushima Lessons Learned: UK Action Plan Published." December 31, 2012. http://news.onr.org.uk/2012/12/fukushima-lessons-learned-uk-action-plan-published/.

———. "Generic Design Assessment (GDA) of New Nuclear Power Stations." http://www.onr.org.uk/new-reactors/.

———. "A Guide to Nuclear Regulation in the UK." May 2016. http://www.onr.org.uk/documents/a-guide-to-nuclear-regulation-in-the-uk.pdf.

———. "IAEA Safeguards in the UK." July 5, 2016. http://www.onr.org.uk/safeguards/iaeauk.htm.

———. *Licensing Nuclear Installations.* 4th edition: January 2015. http://www.onr.org.uk/licensing-nuclear-installations.pdf.

———. *Office for Nuclear Regulation Annual Report and Accounts 2014/15.* HC 164. 2015. http://www.onr.org.uk/documents/2015/annual-report-2014-15.pdf.

———. *Office for Nuclear Regulation Strategic Plan 2016–2020.* March 2016. http://www.onr.org.uk/documents/2016/strategic-plan-2016-2020.pdf.

———. *ONR Review of NGL DNB PLEX Technical Overview Report.* ONR-CNRP-PAR-14-008 rev 0. August 2014. http://www.onr.org.uk/pars/2014/dungeness-b-14-008.pdf.

———. "ONR Staff Recruitment." February 18, 2016. http://news.onr.org.uk/2016/02/onr-staff-recruitment/.

———. *Periodic Safety Reviews (PSR).* NS-TAST-GD-050 rev. 4. April 2013. http://www.onr.org.uk/operational/tech_asst_guides/ns-tast-gd-050.pdf.

———. *Regulation Matters.* July 2016. http://www.onr.org.uk/documents/2016/regulation-matters-july-2016.pdf.

———. *Safety Assessment Principles for Nuclear Facilities.* Rev. 0. 2014. http://www.onr.org.uk/saps/saps2014.pdf.

———. *Security Assessment Principles for the Civil Nuclear Industry.* 2017 Edition, Version 0. http://www.onr.org.uk/syaps/security-assessment-principles-2017.pdf.

———. *Summary Report of the Step 3 Generic Design Assessment (GDA) of Hitachi-GE Nuclear Energy's UK Advanced Boiling Water Reactor (UK ABWR).* October 2015. http://www.onr.org.uk/new-reactors/uk-abwr/reports/step3/uk-abwr-step-3-summary-report.pdf.

———. "What Are Nuclear Safeguards?" March 3, 2016. http://www.onr.org.uk/safeguards/what.htm.

ONR (Office for Nuclear Regulation), et al. *Assessing New Nuclear Reactor Designs: Generic Design Assessment Quarterly Report February 2016—April 2016.* http://www.onr.org.uk/new-reactors/reports/gda-quarterly-report-feb16-apr16.pdf.

The Open University. "National Skills Academy for Nuclear Welcomes The Open University as Their First Higher Education Associate Member." February 11, 2009. http://www3.open.ac.uk/media/fullstory.aspx?id=15602.

———. "Nuclear Energy." 2016. http://www.open.ac.uk/postgraduate/research-degrees/topic/nuclear-energy.

———. "Unclear About Nuclear?" OpenLearn course, 2016. http://www.open.edu/openlearn/science-maths-technology/science/unclear-about-nuclear/content-section-3.4.

Pacala, S., and R. Socolow. "Stabilization Wedges: Solving the Climate Problem for the Next 50 Years with Current Technologies." *Science* 305.5686 (August 13, 2004) 968–72. http://science.sciencemag.org/content/305/5686/968.full

Pagnamenta, Robin. "Ageing Nuclear Plants to Keep Lights on for Years: Safety Fears Rise as Reactors Allowed to Operate Beyond Lifespan." *Times*, February 17, 2016. http://www.thetimes.co.uk/tto/business/industries/utilities/article4692218.ece.

Palmer, Jason, et al. *Warm and Green: Achieving Affordable, Low Carbon Energy While Reducing Impacts on the Countryside.* Cambridge Architectural Research and Anglia Ruskin University, for Campaign to Protect Rural England, April 2015. http://www.cpre.org.uk/resources/energy-and-waste/climate-change-and-energy/item/3903-warm-and-green.

Park, Soo-Ho, et al. "Can Renewable Energy Replace Nuclear Power in Korea? An Economic Valuation Analysis." *Nuclear Engineering and Technology* 48.2 (April 2016) 559–71. http://www.sciencedirect.com/science/article/pii/S1738573316000115.

Parker, David J., et al. "Life Cycle Greenhouse Gas Emissions from Uranium Mining and Milling in Canada." *Environmental Science and Technology* 50.17 (2016) 9746–53. http://pubs.acs.org/doi/abs/10.1021/acs.est.5b06072.

BIBLIOGRAPHY

Parthasarathy, K. S. "The Fallacy of Nuclear Power Being Too Cheap to Meter." *News Minute*, February 19, 2015. http://www.thenewsminute.com/technologies/511.

Patenaude, Bill. "On the Ethics of Energy." Forum on Religion and Ecology at Yale, June 10, 2015. http://fore.yale.edu/news/item/on-the-ethics-of-energy/.

Perko, Tanja. "Radiation Risk Perception: A Discrepancy between the Experts and the General Population." *Journal of Environmental Radioactivity* 133 (July 2014) 86–91. http://www.sciencedirect.com/science/article/pii/S0265931X13000945.

Peterson, Per F. "Spent Fuel is Not the Problem." *Proceedings of the IEEE* 105.3 (March 2017) 411–14. http://ieeexplore.ieee.org/stamp/stamp.jsp?arnumber=7857853.

PHE (Public Health England). "Radon at a Glance." http://www.ukradon.org/information/radonataglance.

———. "Radon, What Can I Do?" Video. February 6, 2015. https://www.youtube.com/watch?v=zI6FRrA23cE.

———. "The Risks to Your Health from Radon." http://www.ukradon.org/information/risks.

Poinssot, Ch., et al. "Assessment of the Environmental Footprint of Nuclear Energy Systems. Comparison Between Closed and Open Fuel Cycles." *Energy* 69 (May 2014) 199–211. http://www.sciencedirect.com/science/article/pii/S0360544214002035.

Pollard, Clare, et al. "Taking Nuclear Skills Development to the Next Level." *Nuclear Future* 12.1 (January/February 2016) 30–33.

Poortinga, Wouter, et al. *Public Perceptions of Nuclear Power, Climate Change and Energy Options in Britain: Summary Findings of a Survey Conducted during October and November 2005*. Understanding Risk Working Paper 06-02. Centre for Environmental Risk, 2006. http://psych.cf.ac.uk/understandingrisk/docs/survey_2005.pdf.

Posiva. "First Excavation Works for Posiva's Final Disposal Facility to Begin - YIT as Contractor." Press release, November 29, 2016. http://www.posiva.fi/en/media/press_releases/first_excavation_works_for_posivas_final_disposal_facility_to_begin_-_yit_as_contractor.3300.news#.WMgsJoXXIcA.

Prance, Ghillean. *Thinking About . . . Creation Care: Is It the Responsibility of Christians to Care for the Environment?* Thinking About . . . leaflet series. Christians in Science, 2016. http://www.cis.org.uk/resources/thinking/.

Prescott, Chris. *Oxford Study Science Dictionary*. Special ed. Oxford: Oxford University Press, 2013.

Prüss-Ustün A., et al. *Preventing Disease Through Healthy Environments: A Global Assessment of the Burden of Disease from Environmental Risks*. World Health Organization, 2016. http://apps.who.int/iris/bitstream/10665/204585/1/9789241565196_eng.pdf.

Qvist, Staffan A., and Barry W. Brook. "Environmental and Health Impacts of a Policy to Phase Out Nuclear Power in Sweden." *Energy Policy* 84 (2015) 1–10. http://bravenewclimate.com/2015/05/05/environmental-and-health-impacts-of-a-policy-to-phase-out-nuclear-power-in-sweden/.

Read, D., et al. "Background in the Context of Land Contaminated with Naturally Occurring Radioactive Material." *Journal of Radiological Protection* 33.2 (2013) 367–80. http://iopscience.iop.org/article/10.1088/0952-4746/33/2/367.

Report of the Tribunal Appointed to Inquire into the Disaster at Aberfan on October 21st, 1966. London: HMSO, 1967. http://www.dmm.org.uk/ukreport/553-01.htm.

Rhodes, Richard. *The Making of the Atomic Bomb*. New York: Simon & Schuster, 1986.

BIBLIOGRAPHY

Roberts, John, and Robin Grimes. "2016 Nuclear Academics Discussion Meeting." *Nuclear Future* 13.1 (January/February 2017) 52.

Rodger, Robert, et al. "The Nuclear 3S – Safety, Security And Safeguards: An Integrated Approach." *Nuclear Future* 13.1 (January/February 2017) 26–31.

Roser-Renouf, C., et al. *Faith, Morality and the Environment: Portraits of Global Warming's Six Americas*. Yale University and George Mason University. New Haven, CT: Yale Program on Climate Change Communication, 2016. http://climatecommunication. yale.edu/wp-content/uploads/2016/01/Faith-Morality-Six-Americas.pdf.

Roxburgh, Al. "Moving from Gloom to Action." *Mission Catalyst* 3 (2014) 14. http://www.bmsworldmission.org/engagecatalyst/mission-catalyst-climate-change-is-boring/moving-gloom-action.

Royal Academy of Engineering. *A Critical Time for UK Energy Policy: What Must Be Done Now to Deliver the UK's Future Energy System*. Report for the Council for Science and Technology. October 2015. http://www.raeng.org.uk/publications/reports/a-critical-time-for-uk-energy-policy.

Royal Society, and Royal Academy of Engineering. *Nuclear Energy: The Future Climate*. 11/99. June 1999. https://royalsociety.org/~/media/Royal_Society_Content/policy/publications/1999/10087-Summary.pdf.

RPII (Radiological Protection Institute of Ireland) and HSE (Health Service Executive). *Radon Gas in Ireland: Joint Position Statement by the Radiological Protection Institute of Ireland and the Health Service Executive*. April 2010. http://www.epa.ie/pubs/reports/radiation/RPII_Radon_Ireland_Joint_HSE_10.pdf.

Rudd, Amber. "Amber Rudd's Speech on a New Direction for UK Energy Policy." Transcript of speech delivered at the Institution of Civil Engineers, London, November 18, 2015. Department of Energy and Climate Change, November 18, 2015. https://www.gov.uk/government/speeches/amber-rudds-speech-on-a-new-direction-for-uk-energy-policy.

Russell, Colin A. *The Earth, Humanity, and God*. Templeton Lectures, 1993. London: UCL, 1994.

RWM (Radioactive Waste Management). "Public Consultation on National Geological Screening." April 21, 2016. https://www.gov.uk/government/consultations/public-consultation-on-national-geological-screening.

Sacks, Bill, et al. "Epidemiology Without Biology: False Paradigms, Unfounded Assumptions, and Specious Statistics in Radiation Science." *Biological Theory*, June 17, 2016. http://www.thesciencecouncil.com/pdfs/Epi%20Without%20Biology%20(BT).pdf.

Sandman, Peter M. "Risk Communication: Facing Public Outrage." *EPA Journal*, November 1987, 21–22. http://www.psandman.com/articles/facing.htm.

Schauer, David A. *Ionizing Radiation Exposure of the Population of the United States*. Presentation on NCRP Report No. 160. 2012. http://ncrponline.org/wp-content/themes/ncrp/PDFs/DAS_DDM2_Athens_4-2012.pdf.

Schleussner, Carl-Friedrich, et al. *Corrigendum to* "Differential Climate Impacts for Policy-Relevant Limits to Global Warming: The Case of 1.5 °C and 2 °C" published in Earth Syst. Dynam., 7, 327–351, 2016. http://www.earth-syst-dynam.net/7/327/2016/esd-7-327-2016-corrigendum.pdf.

———. "Differential Climate Impacts for Policy-Relevant Limits to Global Warming: The Case of 1.5 °C and 2 °C." *Earth System Dynamics* 7.2 (2016) 327–51. http://www.earth-syst-dynam.net/7/327/2016/. doi:10.5194/esd-7-327-2016.

Sciencewise. *New Nuclear Power Stations – Improving Public Involvement in Reactor Design Assessments.* March 2016. http://www.sciencewise-erc.org.uk/cms/assets/Uploads/SWiseGDACSv2.pdf.

Scott, Malcolm, and David Johnson. *Science Matters: Nuclear Power.* 2nd ed. Milton Keynes: The Open University, 1997.

Scott, Tom. "Meeting the Skills Challenge in Nuclear Science and Engineering." Promotional feature. *Nuclear Future* 11.2 (March/April 2015) 34–35.

Scottish Government. "Business, Industry & Energy: Get the Facts: FAQs." September 22, 2016. http://www.gov.scot/Topics/Business-Industry/Energy/Facts/faqs.

———. "The Scottish Government's Response to the UK Government Consultation on the 'Future of Nuclear Power.'" October 9, 2007. http://www.gov.scot/Publications/2007/10/Nuclear.

Sentamu, John. "Kiribati: Living in the Eye of the Climate Change Storm." September 3, 2015. http://www.archbishopofyork.org/articles.php/3312/kiribati-living-in-the-eye-of-the-climate-change-storm.

Shea, Daniel, and Kristy Hartman. *State Options to Keep Nuclear in the Energy Mix.* National Conference of State Legislatures, January 2017. http://www.ncsl.org/Portals/1/Documents/energy/StateOptions_NuclearPower_f02_WEB.pdf.

Shih, Yi-Hsuan, et al. "Socioeconomic Costs of Replacing Nuclear Power with Fossil and Renewable Energy in Taiwan." *Energy* 111 (2016) 369–81. https://www.researchgate.net/publication/306091978_Socioeconomic_costs_of_replacing_nuclear_power_with_fossil_and_renewable_energy_in_Taiwan.

Silin, Maxim. "Layer-Based Data Security in the Real-Time Environment." *Nuclear Future* 9.3 (May/June 2013) 59–63.

Simms, Helen. "Naturally Occurring Radioactive Material." *Nuclear Future* 11.1 (January/February 2015) 28–31.

Slovic, Paul. "Perception of Risk and the Future of Nuclear Power." Paper presented for Session 6, "Public Preferences and Risk Perceptions," of the First MIT International Conference on the Next Generation of Nuclear Power Technology, Cambridge, MA, October 4–5, 1990. International Nuclear Information System (INIS) vol. 37, iss. 31. INIS-XA-N--19. http://www.iaea.org/inis/collection/nclcollectionstore/_public/37/073/37073972.pdf.

SNP (Scottish National Party). *Re-Elect: SNP Manifesto 2016.* https://d3n8a8pro7vhmx.cloudfront.net/thesnp/pages/5540/attachments/original/1461753756/SNP_Manifesto2016-accesible.pdf?1461753756.

———. *Stronger for Scotland.* Manifesto 2015. http://votesnp.com/docs/manifesto.pdf.

Song, Danrong, et al. "Small Is Beautiful: ACP100 Small Modular Reactor Research and Development in China." *Nuclear Future* 12.4 (July/August 2016) 44–51.

Spencer, Nick, and Robert White. *Christianity, Climate Change and Sustainable Living.* London: SPCK, 2007.

SRTP (Society, Religion and Technology Project). *Nuclear Weapons.* Church of Scotland Church and Society Council. http://www.srtp.org.uk/assets/uploads/Nuclear_weapons_866_0714.pdf.

———. "What Future for Nuclear Power?" April 14, 2010. http://www.srtp.org.uk/srtp/view_article/what_future_for_nuclear_power.

Stainsby, Richard. "Advanced Fuel Cycle Research and Development." *Nuclear Future* 10.1 (January/February 2014) 31–35.

BIBLIOGRAPHY

Starling, Rosie. "IEA: World Energy Investment 2016." Hydrocarbon Engineering, September 14, 2016. https://www.energyglobal.com/downstream/refining/14092016/iea-world-energy-investment-2016-4077/.

Steare, Roger, et al. *Managers and Their MoralDNA: Better Values, Better Business.* Chartered Management Institute, March 2014. http://www.managers.org.uk/insights/research/current-research/2014/march/managers-and-their-moraldna.

Stern, Nicholas. *A Blueprint for a Safer Planet: How to Manage Climate Change and Create a New Era of Progress and Prosperity.* London: Bodley Head, 2009.

———. *The Economics of Climate Change: The Stern Review.* Cambridge: Cambridge University Press, 2007.

Stewart, Alex G., et al. "Real or Illusory? Case Studies on the Public Perception of Environmental Health Risks in the North West of England." *International Journal of Environmental Research and Public Health* 7 (2010) 1153–73. http://www.ncbi.nlm.nih.gov/pmc/articles/PMC2872300/pdf/ijerph-07-01153.pdf.

Stone, Robert, director. *Pandora's Promise.* Film. CNN Films, 2013. http://pandoraspromise.com.

———. "The Education of an Environmentalist: How an Award-Winning Filmmaker Who Created the Definitive Earth Day Documentary Learned to Love Nuclear Power in an Age of Global Warming." *Scientific American Guest Blog,* April 21, 2016. http://blogs.scientificamerican.com/guest-blog/the-education-of-an-environmentalist/.

Storkey, Elaine. "The Environment and the Developing World." In *Wisdom, Science and the Scriptures: Essays in Honour of Ernest Lucas,* edited by Stephen Finamore and John Weaver, 118–32. Oxford: Centre for Baptist History and Heritage, and Bristol Baptist College, 2012.

Stover, Dawn. "Kerry Emanuel: A Climate Scientist for Nuclear Energy." Interview by Dawn Stover. *Bulletin of the Atomic Scientists* 73.1 (2017) 7–12. Published online December 14, 2016. Special Issue: Should Nuclear Power Be a Major Part of the World's Response to Climate Change. http://www.tandfonline.com/doi/full/10.1080/00963402.2016.1264205.

———. "More Megatons to Megawatts." *Bulletin of the Atomic Scientists,* February 21, 2014. http://thebulletin.org/more-megatons-megawatts.

———. "Nuclear vs. Renewables: Divided They Fall." *Bulletin of the Atomic Scientists,* January 30, 2014. http://thebulletin.org/nuclear-vs-renewables-divided-they-fall.

Taebi, Behnam. *Good Governance of Risky Technology: Bridging the Acceptance-Acceptability-Gap.* Presented at the Second North American Workshop on the Ethical Dimensions of the System of Radiological Protection, held at the Belfer Center for Science and International Affairs, Harvard University, Cambridge, Massachusetts, March 10–12, 2015. http://www.icrp.org/docs/7/4.%20Bridging%20the%20gap%20between%20public%20acceptance%20and%20ethical%20acceptability%20-Taebi.pdf.

———. "The Morally Desirable Option for Nuclear Power Production." *Philosophy and Technology* 24.2 (2011) 169–92. http://link.springer.com/article/10.1007%2Fs13347-011-0022-y.

Taebi, Behnam, and Sabine Roeser. "The Ethics of Nuclear Energy: An Introduction." In *The Ethics of Nuclear Energy: Risk, Justice, and Democracy in the Post-Fukushima Era,* edited by Taebi Behnam and Sabine Roeser, 1–14 Cambridge: Cambridge University Press, 2015. http://assets.cambridge.org/97811070/54844/excerpt/9781107054844_excerpt.pdf.

BIBLIOGRAPHY

Teske, Sven, et al. *100% Renewable Energy for All: Energy [R]evolution Scenario – A Sustainable World Energy Outlook 2015. Executive Summary*. 5th ed. Greenpeace, September 2015. http://www.greenpeace.org/international/Global/international/publications/climate/2015/Energy-Revolution-2015-Summary.pdf.

Thomas, Philip. "Why We Need a New Science of Safety." *The Conversation*, July 20, 2016. http://theconversation.com/why-we-need-a-new-science-of-safety-58436.

3KQ (Three Key Questions). *New Nuclear Power Stations: Improving Public Involvement in Reactor Design Assessments*. Public Dialogue Report prepared for the EA, ONR, and NRW. August 2015. http://www.sciencewise-erc.org.uk/cms/assets/Uploads/GDA-dialogue-report-August-2015-FINAL.pdf.

Tindale, Stephen. "UK Climate and Energy Policy: Small Steps Forward, Large Steps Backwards." *Climate Answers* (blog), December 7, 2015. http://climateanswers.info/2015/12/7-december-2015-uk-climate-and-energy-policy-small-steps-forward-large-steps-backwards/.

———. "Well Done Amber Rudd." *Climate Answers* (blog), November 19, 2015. http://climateanswers.info/2015/11/19-november-2015-well-done-amber-rudd/.

Tomlinson, Hugh. "'Banned' Firework Competition." *Times*, April 11, 2016, 32.

Trades Union Congress. *Powering Ahead: How UK Industry Can Match Europe's Environmental Leaders*. Economic Report Series. July 2016. https://www.tuc.org.uk/sites/default/files/Powering_Ahead_Report.pdf.

———. "Unions 4 Climate Action." June 4, 2014. https://www.tuc.org.uk/unions4climateaction.

Trainer, Ted. "Can Renewables etc. Solve the Greenhouse Problem? The Negative Case." *Energy Policy* 38 (2010) 4107–14. http://jayhanson.us/_Energy/TrainerRenewables.pdf.

Travis, Karl. "Deep Borehole Disposal." *Nuclear Future* 12.5 (September/October 2016) 32–36.

Tucker, Mary Evelyn, and John Grim. "Integrating Ecology and Justice: The Papal Encyclical." *Quarterly Review of Biology* 91.3 (September 2016) 261–70. http://fore.yale.edu/files/Integrating_Ecology_and_Justice.pdf.

Tyndall Centre for Climate Change Research. *A Review of Research Relevant to New Nuclear Build Power Plants in the UK*. Commissioned by Friends of the Earth. January 2013. https://www.foe.co.uk/sites/default/files/downloads/tyndall_evidence.pdf.

UN (United Nations). "Sustainable Development Goals: 17 Goals to Transform Our World." http://www.un.org/sustainabledevelopment/sustainable-development-goals/.

UNFCCC (United Nations Framework Convention on Climate Change). *Adoption of the Paris Agreement*. Conference of the Parties, 21st session, Paris, November 30—December 11, 2015. FCCC/CP/2015/L.9/rev. 1. December 12, 2015. https://unfccc.int/resource/docs/2015/cop21/eng/l09r01.pdf.

Unionlearn. "TUC Visits New Hinkley Point Learning Centre." January 20, 2016. https://www.unionlearn.org.uk/news/tuc-visits-new-hinkley-point-learning-centre.

United Nations Environment Programme. *Radiation: Effects and Sources*. 2016. http://www.fs-ev.org/fileadmin/user_upload/89_News/Oeff.-Arbeit/Radiation_Effects_and_sources-2016.pdf.

United Nations Information Service. "New UN Study Assesses Radiation Exposure from Electricity Generation Technologies." Press release, UNIS/OUS/366, February 8,

2017. http://www.unscear.org/docs/media/UNSCEAR_2016_report_pressrelease.pdf.

University of Bristol Cabot Institute. "Bristol Named as National College for Nuclear Partner to Combat Future Industry Skills Shortage." March 20, 2015. http://www.bris.ac.uk/cabot/news/2015/bristol-nuclear-partner.html.

———. "Is Nuclear Green?" 2015. http://www.bristol.ac.uk/cabot/events/2015/is-nuclear-green.html.

University of Cumbria. "BIS Announces £15 Million in Government Funding to Set Up the National College for Nuclear (NCfN) Headquartered in Cumbria and Somerset." May 9, 2016. http://my.cumbria.ac.uk/AboutUs/News/Articles/201516/May/NCfN-announcement.aspx.

USNIC (United States Nuclear Infrastructure Council). *USNIC Backend Working Group Issue Brief: Charting a Path Forward.* March 2017. https://media.wix.com/ugd/7607 34_8081B188C2BA4714AF7610CCB0FCF4B5.pdf.

Van Dine, Alexandra, et al. *Outpacing Cyber Threats: Priorities for Cybersecurity at Nuclear Facilities.* Nuclear Threat Initiative, 2016. http://www.nti.org/media/documents/NTI_CyberThreats__FINAL.pdf.

Vidal, John. "Environmental Risks Killing 12.6 Million People, WHO Study Says." *Guardian*, March 15, 2016. http://www.theguardian.com/environment/2016/mar/15/environmental-risks-killing-126-million-people-who-study-says.

Wakeford, Richard. Review of *NCRP Report No. 160: Ionizing Radiation Exposure of the Population of the United States. Journal of Radiological Protection* 29.3 (2009) 465. http://iopscience.iop.org/article/10.1088/0952-4746/29/3/B01.

———. "The Windscale Reactor Accident—50 Years On." *Journal of Radiological Protection* 27.3 (2007) 211–15. http://iopscience.iop.org/article/10.1088/0952-4746/27/3/E02.

Wall, B. F., et al. *Radiation Risks from Medical X-Ray Examinations as a Function of the Age and Sex of the Patient.* HPA-CRE-028. HPA, October 2011. https://www.gov.uk/government/uploads/system/uploads/attachment_data/file/340147/HPA-CRCE-028_for_website.pdf.

Walli, Ron. "ORNL Technique Could Set New Course for Extracting Uranium from Seawater." Oak Ridge National Laboratory, December 17, 2015. https://ornl.gov/news/ornl-technique-could-set-new-course-extracting-uranium-seawater.

Walton, John H. *The Lost World of Genesis One: Ancient Cosmology and the Origins Debate.* Downers Grove, IL: IVP Academic, 2009.

WANO (World Association of Nuclear Operators). *2015 Year-End Highlights Report.* http://www.wano.info/en-gb/library/highlightsreport/Documents/Year%20End%20Highlights%20Report%202015%20Final.pdf.

Watson, S. J., et al. *Ionising Radiation Exposure of the UK Population: 2005 Review.* HPA-RPD-001. HPA, May 2005. https://www.gov.uk/government/publications/ionising-radiation-exposure-of-the-uk-population-2005-review.

WCC (World Council of Churches) Central Committee. "Statement Towards a Nuclear-Free World." WCC, July 7, 2014. http://www.oikoumene.org/en/resources/documents/central-committee/geneva-2014/statement-towards-a-nuclear-free-world.

Weaver, John. *Christianity and Science.* London: SCM, 2010.

———. "Exploring Hope." *Anvil* 29.1 (2013) 25–41. https://www.degruyter.com/view/j/anv.2013.29.issue-1/anv-2013-0003/anv-2013-0003.xml.

BIBLIOGRAPHY

———. "Urgent: Christians and Climate Change." European Baptist Federation, September 7, 2016. http://www.ebf.org/urgent-christians-and-climate-change.

Weaver, John, and Margot R. Hodson, eds. *The Place of Environmental Theology: A Course Guide for Seminaries, Colleges and Universities*. Oxford: Whitley Trust, 2007. http://www.cebts.eu/programme/doc/environmental-theology-course-2007.pdf.

WEC (World Energy Council). "The Road to Resilience: Managing Cyber Risks." September 2016. https://www.worldenergy.org/publications/2016/the-road-to-resilience-managing-cyber-risks/.

———. *World Energy Issues Monitor 2016*. March 2016. https://www.worldenergy.org/publications/2016/world-energy-issues-monitor-2016/.

WEF (World Economic Forum). *The Global Risks Report 2016*. 11th ed. World Economic Forum, 2016. http://reports.weforum.org/global-risks-2016/.

White, Robert S. *The Age of the Earth*. Faraday Paper 8. Faraday Institute for Science and Religion, St. Edmund's College, University of Cambridge, 2007. http://www.faraday.st-edmunds.cam.ac.uk/resources/Faraday%20Papers/Faraday%20Paper%208%20White_EN.pdf.

———, ed. *Creation in Crisis: Christian Perspectives on Sustainability*. London: SPCK, 2009.

The White House (US). "The Nuclear Security Summits: Securing the World from Nuclear Terrorism." Fact sheet. March 29, 2016. https://www.whitehouse.gov/the-press-office/2016/03/29/fact-sheet-nuclear-security-summits-securing-world-nuclear-terrorism.

WHO (World Health Organization). "Alcohol." Fact sheet. January 2015. http://www.who.int/mediacentre/factsheets/fs349/en/#.

———. *Communicating Radiation Risks in Paediatric Imaging: Information to Support Healthcare Discussions About Benefit and Risk*. 2016. http://www.who.int/ionizing_radiation/pub_meet/radiation-risks-paediatric-imaging/en/.

———. "FAQs: Fukushima Five Years On." 2016. http://www.who.int/ionizing_radiation/a_e/fukushima/faqs-fukushima/en/.

———. "Global Strategy to Reduce the Harmful Use of Alcohol." 2010. http://www.who.int/substance_abuse/activities/gsrhua/en/.

———. "Ionizing Radiation, Health Effects and Protective Measures." Fact sheet. April 2016. http://www.who.int/mediacentre/factsheets/fs371/en/.

———. "Tobacco." Fact sheet. June 2016. http://www.who.int/mediacentre/factsheets/fs339/en/.

Wikipedia. "Swords to Ploughshares." Modified April 6, 2017. https://en.wikipedia.org/wiki/Swords_to_ploughshares.

Wilkinson, W. L. "Management of the UK Plutonium Stockpile: The Economic Case for Burning as MOX in New PWRs." *Interdisciplinary Science Review* 26.4 (2001) 303–6. http://www.tandfonline.com/doi/pdf/10.1179/isr.2001.26.4.303.

Williams, Laurence. "Nuclear Safety, Security and Safeguards: The Need for Greater Integration." *Nuclear Future* 9.3 (May/June 2013) 33–41.

WIPP (Waste Isolation Pilot Plant). "About WIPP." February 25, 2016. http://www.wipp.energy.gov/wipprecovery/about.html.

———. "Plans and Reports." February 25, 2016. http://www.wipp.energy.gov/wipprecovery/plans_reports.html.

———. "Status Update on WIPP Restart." WIPP Update, July 15, 2016. http://www.wipp.energy.gov/Special/WIPP%20Update%207_15_16%20.pdf.

———. "WIPP Recovery Plan." http://www.wipp.energy.gov/wipprecovery/path_forward.html.

———. "WIPP Update: April 10, 2017 – WIPP Receives First Shipment Since Reopening." http://www.wipp.energy.gov/Special/WIPP%20Update%204_10_17.pdf.

Wiseman, Jennifer. "In Support of the Pope's Statement on Climate Change." *Huffington Post*, July 17, 2016. http://www.huffingtonpost.com/jennifer-wiseman/in-support-of-the-popes-statement-on-climate-change_b_7806614.html.

Witty, Andrew. *Encouraging a British Invention Revolution: Sir Andrew Witty's Review of Universities and Growth*. Department for Business, Innovation and Skills, October 2013. https://www.gov.uk/government/uploads/system/uploads/attachment_data/file/291911/bis-13-1241-encouraging-a-british-invention-revolution-andrew-witty-review-R1.pdf.

WNA (World Nuclear Association). "Advanced Nuclear Power Reactors." September 2016. http://world-nuclear.org/information-library/nuclear-fuel-cycle/nuclear-power-reactors/advanced-nuclear-power-reactors.aspx.

———. "Charter of Ethics." 2016. http://www.world-nuclear.org/our-association/membership/charter-of-ethics.aspx.

———. "Fast Neutron Reactors." March 2017. http://www.world-nuclear.org/information-library/current-and-future-generation/fast-neutron-reactors.aspx.

———. "Generation IV Nuclear Reactors." July 2016. http://www.world-nuclear.org/information-library/nuclear-fuel-cycle/nuclear-power-reactors/generation-iv-nuclear-reactors.aspx.

———. "Nuclear Fusion Power." May 2016. http://world-nuclear.org/information-library/current-and-future-generation/nuclear-fusion-power.aspx.

———. "Nuclear Power in Japan." April 2016. http://www.world-nuclear.org/information-library/country-profiles/countries-g-n/japan-nuclear-power.aspx.

———. "Nuclear Power in Sweden." September 2016. http://www.world-nuclear.org/information-library/country-profiles/countries-o-s/sweden.aspx.

———. "Nuclear Power in the United Kingdom." October 2016. http://www.world-nuclear.org/information-library/country-profiles/countries-t-z/united-kingdom.aspx.

———. *Nuclear Power Reactor Characteristics*. 2016/17 Pocket Guide. August 2016. http://www.world-nuclear.org/getmedia/80f869be-32c8-46e7-802d-eb4452939ec5/Pocket-Guide-Reactors.pdf.aspx.

———. "Nuclear-Powered Ships." January 2017. http://www.world-nuclear.org/information-library/non-power-nuclear-applications/transport/nuclear-powered-ships.aspx.

———. "Research Reactors." September 2016. http://www.world-nuclear.org/information-library/non-power-nuclear-applications/radioisotopes-research/research-reactors.aspx.

———. "Royal Commission's Conclusions Create Middle-Ground in the Nuclear Waste Discourse." May 9, 2016. http://www.world-nuclear.org/press/press-statements/royal-commission's-conclusions-create-middle-groun.aspx.

———. "Safeguards to Prevent Nuclear Proliferation." April 2016. http://www.world-nuclear.org/information-library/safety-and-security/non-proliferation/safeguards-to-prevent-nuclear-proliferation.aspx.

———. "Small Nuclear Power Reactors." February 27, 2017. http://www.world-nuclear.org/information-library/nuclear-fuel-cycle/nuclear-power-reactors/small-nuclear-power-reactors.aspx.

———. "Supply of Uranium." December 2016. http://www.world-nuclear.org/info/Nuclear-fuel-cycle/Uranium-Resources/Supply-of-Uranium/.

———. "Thorium." September 2015. http://www.world-nuclear.org/information-library/current-and-future-generation/thorium.aspx.

———. *World Nuclear Performance Report 2016.* Report 2016/001. June 2016. http://www.world-nuclear.org/our-association/publications/online-reports/world-nuclear-performance-report-2016.aspx.

WNN (*World Nuclear News*). "Academy Highlights Contradiction in French Policy." April 21, 2017. http://www.world-nuclear-news.org/EE-Academy-highlights-contradiction-in-French-policy-2104174.html.

———. "AP1000 Design Completes UK Regulatory Assessment." March 30, 2017. http://www.world-nuclear-news.org/RS-AP1000-design-completes-UK-regulatory-assessment-30031701.html.

———. "British Reactor Takes Record for Longest Continuous Operation." August 2, 2016. http://www.world-nuclear-news.org/C-British-reactor-takes-record-for-longest-continuous-operation-0208164.html.

———. "British University Unveils 'Diamond' Nuclear-Powered Battery." November 28, 2016. http://www.world-nuclear-news.org/WR-British-university-unveils-diamond-nuclear-powered-battery-28111602.html.

———. "Climate Target 'Very Difficult' Without Nuclear, Says IEA Chief Economist." December 6, 2016. http://www.world-nuclear-news.org/EE-Climate-target-very-difficult-without-nuclear-says-IEA-chief-economist-06121601.html.

———. "Construction to Start on Finnish Repository." November 29, 2016. http://www.world-nuclear-news.org/WS-Construction-to-start-on-Finnish-repository-2911164.html.

———. "Court Backs German Utilities' Compensation Claim." December 6, 2016. http://www.world-nuclear-news.org/NP-Court-backs-German-utilities-compensation-claim-0612144.html.

———. "Cyber Risk Must Be Managed, Says World Energy Council." October 5, 2016. http://www.world-nuclear-news.org/ON-Cyber-risk-must-be-managed-says-World-Energy-Council-05101602.html.

———. "EDF Energy Extends Lives of UK AGR Plants." February 16, 2016. http://www.world-nuclear-news.org/C-EDF-Energy-extends-lives-of-UK-AGR-plants-1602164.html.

———. "EIA Sees Strong Growth in Nuclear Generation to 2040." May 12, 2016. http://www.world-nuclear-news.org/EE-EIA-sees-strong-growth-in-nuclear-generation-to-2040–1205164.html.

———. "Europe Needs to Revise Nuclear Strategy, Says Committee." September 26, 2016. http://www.world-nuclear-news.org/NP-Europe-needs-to-revise-nuclear-strategy-says-committee-2609164.html.

———. "First Criticality for Watts Bar 2." May 24, 2016. http://www.world-nuclear-news.org/NN-First-criticality-for-Watts-Bar-2-2405167.html.

———. "Horizon Applies for Wylfa Newydd Site Licence." April 4, 2017. http://www.world-nuclear-news.org/RS-Horizon-applies-for-Wylfa-Newydd-site-licence-04041701.html.

———. "Ikata 3 Back in Commercial Operation." September 7, 2016. http://www.world-nuclear-news.org/C-Ikata-3-back-in-commercial-operation-0709165.html.

———. "Illinois Energy Bill Becomes Law." December 8, 2016. http://www.world-nuclear-news.org/NP-Illinois-energy-bill-becomes-law-0812168.html.

———. "India Creates Medical Supplies from Nuclear Waste." February 20, 2017. http://www.world-nuclear-news.org/ON-India-creates-medical-supplies-from-nuclear-waste-2002171.html.

———. "Japanese Institute Sees 19 Reactor Restarts by March 2018." July 28, 2016. http://www.world-nuclear-news.org/NP-Japanese-institute-sees-19-reactor-restarts-by-March-2018-2807164.html.

———. "Low Emissions from Canadian Uranium Mining." September 9, 2016. http://www.world-nuclear-news.org/EE-Low-emissions-from-Canadian-uranium-mining-0906167.html.

———. "Near-Surface Final Waste Repository Launched in Russia." December 15, 2016. http://www.world-nuclear-news.org/WR-Near-surface-final-waste-repository-launched-in-Russia-15121601.html.

———. "New Mexico Used Fuel Project Put to Regulators." March 31, 2017. http://www.world-nuclear-news.org/WR-New-Mexico-used-fuel-project-put-to-regulators-3104171.aspx.

———. "New York Approves Clean Energy Standard." August 2, 2016. http://www.world-nuclear-news.org/NP-New-York-approves-Clean-Energy-Standard-0208167.html.

———. "Nuclear Industry Needs to Improve Communication." April 14, 2016. http://www.world-nuclear-news.org/RS-Nuclear-industry-needs-to-improve-communication-14041601.html.

———. "Nuclear Plays 'Vital' Role in UK Economy, Statistics Show." April 11, 2016. http://www.world-nuclear-news.org/EE-Nuclear-plays-vital-role-in-UK-economy-statistics-show-11041601.html.

———. "Nuclear Ship Takes Olympic Flame to North Pole." October 25, 2013. http://www.world-nuclear-news.org/ON-Nuclear_ship_takes_Olympic_flame_to_North_Pole-2510135.html.

———. "Nuclear Vital to Challenge of Climate Change." November 18, 2016. http://www.world-nuclear-news.org/EE-Nuclear-vital-to-challenge-of-climate-change-18111602.html.

———. "Nuclear 'Vital' to Sustainable Development Goals." September 29, 2016. http://www.world-nuclear-news.org/EE-Nuclear-vital-to-Sustainable-Development-Goals-29091601.html.

———. "Nuclear's Role in UK's Low-Carbon, Industrial Strategies." January 25, 2017. http://www.world-nuclear-news.org/NP-Nuclears-role-in-UKs-low-carbon-industrial-strategies-25011701.html.

———. "NuScale Makes History with SMR Design Application." January 13, 2017. http://www.world-nuclear-news.org/NN-NuScale-makes-history-with-SMR-design-application-13011701.html.

———. "Record 940 Days of Continuous Operation for Heysham Unit." September 16, 2016. http://www.world-nuclear-news.org/C-Record-940-days-of-continuous-operation-for-Heysham-unit-1609164.html.

———. "Red Book Sees Production Capacity as Future Uranium Challenge." December 5, 2016. http://www.world-nuclear-news.org/UF-Red-Book-sees-production-capacity-as-future-uranium-challenge-0512167.html.

BIBLIOGRAPHY

———. "Rolls-Royce Names Partners for UK SMR." January 9, 2017. http://www.world-nuclear-news.org/NN-Rolls-Royce-names-partners-for-UK-SMR-09011701.html.

———. "Russia Hails Progress with Americium in Fast Reactor Research." December 22, 2015. http://www.world-nuclear-news.org/WR-Russia-hails-progress-with-americium-in-fast-reactor-research-22121501.html.

———. "Russia Suspends Plutonium Agreement with USA." October 4. 2016. http://www.world-nuclear-news.org/NP-Russia-suspends-plutonium-agreement-with-USA-04101601.html.

———. "Russian Fast Reactor Reaches Full Power." August 17, 2016. http://www.world-nuclear-news.org/NN-Russian-fast-reactor-reaches-full-power-1708165.html.

———. "Russia's BN-800 Unit Enters Commercial Operation." November 1, 2016. http://www.world-nuclear-news.org/NN-Russias-BN-800-unit-enters-commercial-operation-01111602.html.

———. "Shikoku Decides Against Restarting Ikata 1." March 29, 2016. http://www.world-nuclear-news.org/C-Shikoku-decides-against-restarting-Ikata-1-2903164.html.

———. "Study Confirms NuScale Reactor's MOX Capability." January 21, 2016. http://www.world-nuclear-news.org/WR-Study-confirms-NuScale-reactors-MOX-capability-2101164.html.

———. "Summit Urges Action to Preserve US Nuclear Reactors." May 24, 2016. http://www.world-nuclear-news.org/NP-Summit-urges-action-to-preserve-US-nuclear-reactors-2405169.html.

———. "Sweden Abolishes Nuclear Tax." June 10, 2016. http://www.world-nuclear-news.org/NP-Sweden-abolishes-nuclear-tax-1006169.html.

———. "UK Businesses Plan for Global SMR Market." March 2, 2017. http://www.world-nuclear-news.org/NP-UK-businesses-plan-for-global-SMR-market-0203177.html.

———. "UK Civil Nuclear Job Count Rises by 2000." September 14, 2016. http://www.world-nuclear-news.org/C-UK-civil-nuclear-job-count-rises-by-2000-1409164.html.

———. "UK Considers How to Use Small Reactor Opportunity." October 18, 2016. http://www.world-nuclear-news.org/NN-UK-considers-how-to-use-small-reactor-opportunity-1910161.html.

———. "UK Nuclear's Future in Government Hands, Say Reports." May 2, 2017. http://www.world-nuclear-news.org/NP-UK-nuclear-industrys-future-in-government-hands-say-reports-0205174.html.

———. "UK Parliamentary Hearings Assess Brexit Impact." August 3, 2016. http://www.world-nuclear-news.org/NP-UK-Parliamentary-hearings-assess-Brexit-impact-03081601.html.

———. "UK Parties Make Scant Reference to Nuclear Power." May 19, 2017. http://www.world-nuclear-news.org/NP-UK-parties-make-scant-reference-to-nuclear-power-19051701.html.

———. "UK Sets Out Decommissioning Plans to 2020." April 10, 2017. http://www.world-nuclear-news.org/WR-UK-sets-out-decommissioning-plans-to-2020-10041702.html.

———. "UK to Start Regulatory Assessment of Chinese Design." January 10, 2017. http://www.world-nuclear-news.org/RS-UK-to-start-regulatory-assessment-of-Chinese-design-10011701.html.

———. "UNSCEAR Studies Radiation Exposure from Electricity." February 9, 2017. http://www.world-nuclear-news.org/EE-UNSCEAR-studies-radiation-exposure-from-electricity-0902174.html.

———. "US Administration Urged to Address Waste Issue." May 2, 2017. http://www.world-nuclear-news.org/WR-US-administration-urged-to-address-waste-issue-0205177.html.

———. "USA Sets Out Nuclear Security Strategy." April 6, 2016. http://www.world-nuclear-news.org/NP-NNSA-annual-report-sets-out-strategy-0604167.html.

———. "Waste Shipments to WIPP Expected to Resume Soon." February 16, 2017. http://www.world-nuclear-news.org/WR-Waste-shipments-to-WIPP-expected-to-resume-soon-1602174.html.

———. "Watts Bar 2 Begins Commercial Operation." October 20, 2016. http://www.world-nuclear-news.org/NN-Watts-Bar-2-begins-commercial-operation-2010165.html.

———. "Westinghouse Files for US Bankruptcy Protection." March 29, 2017. http://www.world-nuclear-news.org/C-Westinghouse-files-for-US-bankruptcy-protection-29031702.html.

———. "WIPP Waste Shipments Resume." April 11, 2017. http://www.world-nuclear-news.org/WR-WIPP-waste-shipments-resume-1104177.html.

———. "Wisconsin Lifts Nuclear Moratorium." April 4, 2016. http://www.world-nuclear-news.org/NP-Wisconsin-lifts-nuclear-moratorium-0404168.html.

———. "Worldwide Nuclear Capacity Continues to Grow in 2016." January 3, 2017. http://www.world-nuclear-news.org/NP-Worldwide-nuclear-capacity-continues-to-grow-in-2016-0301175.html.

World Bank, and Institute for Health Metrics and Evaluation. *The Cost of Air Pollution: Strengthening the Economic Case for Action*. World Bank Group, 2016. http://documents.worldbank.org/curated/en/781521473177013155/The-cost-of-air-pollution-strengthening-the-economic-case-for-action.

Wright, N. T. *God in Public: How the Bible Speaks Truth to Power Today*. London: SPCK, 2016.

Xing, Ji., et al. "HPR1000: Advanced Pressurized Water Reactor with Active and Passive Safety." *Engineering* 2.1 (March 2016) 79–87. http://www.sciencedirect.com/science/article/pii/S2095809916301515.

Yeo, Tim. *Five-Point Plan for the UK Nuclear Energy Industry*. New Nuclear Watch Europe, February 27, 2017. http://newnuclearwatch.eu/wp-content/uploads/2017/02/NNWE-5-Point-Plan-for-the-UK-Nuclear-Energy-Industry-FINAL.pdf.

Zeeb, Hajo, and Ferid Shannoun, eds. *WHO Handbook on Indoor Radon: A Public Health Perspective*. WHO, 2009. http://apps.who.int/iris/bitstream/10665/44149/1/9789241547673_eng.pdf.

www.ingramcontent.com/pod-product-compliance
Lightning Source LLC
Chambersburg PA
CBHW051636230426
43669CB00013B/2320